I0064676

Generalized Synchronization and Generalized Consensus of System Arrays

WORLD SCIENTIFIC SERIES ON NONLINEAR SCIENCE

Editor: Leon O. Chua
University of California, Berkeley

Series B. SPECIAL THEME ISSUES AND PROCEEDINGS

Volume 1: Chua's Circuit: A Paradigm for Chaos
Edited by *R. N. Madan*

Volume 2: Complexity and Chaos
Edited by *N. B. Abraham, A. M. Albano, A. Passamante, P. E. Rapp and R. Gilmore*

Volume 3: New Trends in Pattern Formation in Active Nonlinear Media
Edited by *V. Perez-Villar, V. Perez-Munuzuri, C. Perez Garcia and V. I. Krinsky*

Volume 4: Chaos and Nonlinear Mechanics
Edited by *T. Kapitaniak and J. Brindley*

Volume 5: Fluid Physics — Lecture Notes of Summer Schools
Edited by *M. G. Velarde and C. I. Christov*

Volume 6: Dynamics of Nonlinear and Disordered Systems
Edited by *G. Martínez-Mekler and T. H. Seligman*

Volume 7: Chaos in Mesoscopic Systems
Edited by *H. A. Cerdeira and G. Casati*

Volume 8: Thirty Years After Sharkovskiĭ's Theorem: New Perspectives
Edited by *L. Alsedà, F. Balibrea, J. Llibre and M. Misiurewicz*

Volume 9: Discretely-Coupled Dynamical Systems
Edited by *V. Pérez-Muñuzuri, V. Pérez-Villar, L. O. Chua and M. Markus*

Volume 10: Nonlinear Dynamics & Chaos
Edited by *S. Kim, R. P. Behringer, H.-T. Moon and Y. Kuramoto*

Volume 11: Chaos in Circuits and Systems
Edited by *G. Chen and T. Ueta*

Volume 12: Dynamics and Bifurcation of Patterns in Dissipative Systems
Edited by *G. Dangelmayr and I. Oprea*

Volume 13: Modeling and Computations in Dynamical Systems: In Commemoration of the 100th Anniversary of the Birth of John von Neumann
Edited by *E. J. Doedel, G. Domokos and I. G. Kevrekidis*

Volume 14: Dynamics and Control of Hybrid Mechanical Systems
Edited by *G. Leonov, H. Nijmeijer, A. Pogromsky and A. Fradkov*

Volume 15: From Physics to Control Through an Emergent View
Edited by *L. Fortuna, A. Fradkov and M. Frasca*

Volume 16: Frontiers in the Study of Chaotic Dynamical Systems with Open Problems
Edited by *Z. Elhadj and J. C. Sprott*

Volume 17: Advances on Nonlinear Dynamics of Electronic Systems
Edited by *A. Buscarino, L. Fortuna and R. Stoop*

Volume 18: Generalized Synchronization and Generalized Consensus of System Arrays
Edited by *L. Min and G. Chen*

WORLD SCIENTIFIC SERIES ON NONLINEAR SCIENCE Series B Vol. 18

Series Editor: Leon O. Chua

Generalized Synchronization and Generalized Consensus of System Arrays

Lequan Min

University of Science and Technology Beijing, China

Guanrong Chen

City University of Hong Kong, Hong Kong

NEW JERSEY • LONDON • SINGAPORE • BEIJING • SHANGHAI • HONG KONG • TAIPEI • CHENNAI • TOKYO

Published by

World Scientific Publishing Co. Pte. Ltd.
5 Toh Tuck Link, Singapore 596224
USA office: 27 Warren Street, Suite 401-402, Hackensack, NJ 07601
UK office: 57 Shelton Street, Covent Garden, London WC2H 9HE

Library of Congress Control Number: 2020034677

British Library Cataloguing-in-Publication Data
A catalogue record for this book is available from the British Library.

World Scientific Series on Nonlinear Science Series B — Vol. 18
GENERALIZED SYNCHRONIZATION AND GENERALIZED CONSENSUS OF SYSTEM ARRAYS

ISBN 978-981-121-427-1 (hardcover)
ISBN 978-981-121-428-8 (ebook for institutions)
ISBN 978-981-121-429-5 (ebook for individuals)

For any available supplementary material, please visit
https://www.worldscienti ic.com/worldscibooks/10.1142/11664#t=suppl

Typeset by Stallion Press
Email: enquiries@stallionpress.com

Dedicated to the 84th Birthday of Professor Leon O. Chua

Preface

The notion of *synchronization* is classical, which is based on the concept of closeness of the states or the frequencies between two dynamical systems. The earliest study of synchronization led to the discovery of Christian Huygens in 1665 about the perfect synchrony of two pendulum clocks. Since then, many synchronizable systems have been found in various fields of natural and social sciences, engineering and technology, and many theories and methodologies for synchronization analysis and implementation have been developed.

Lately, the notion of synchronization has evolved from that between two coupled systems to that between two coupled arrays of systems, and been generalized to many more phenomena in much wider sense from the classical setting. One typical example is sensor networks, where there are many clusters of sensors for which clock synchronization is essential, and another example is integrated-circuit chips embedded with millions of cells that are working in coordination and synchrony.

From a theoretical perspective, understanding the relationships and interactions between two arrays of nonlinear dynamic systems, either autonomous or non-autonomous, is of fundamental significance. One of such dynamic relationships is synchronization; another is consensus. Roughly speaking, synchronization means that the dynamical behaviors of two system arrays tend to be identical as time evolves, while consensus here means that the difference between the dynamics of the two arrays becomes sufficiently small when the processing time is long enough, where the latter becomes the former if the small difference actually becomes zero.

In more general situations, the above dynamic relationships can be described and understood in a generalized sense, by comparing a certain variant or transform of the dynamics of one array to that of the other array. Such general settings are referred to as generalized synchronization (GS) and generalized consensus (GC), respectively, which will be more precisely defined in the text of the book.

On the other hand, if the evolution time variable of the dynamic systems is continuous, they are described by ordinary differential equations or systems; if the evolution time variable of the dynamic systems is discrete they are described by difference equations or systems.

Moreover, the connections between the two arrays of dynamical systems can be either directional or non-directional, which describes the signal flows from one array to another.

Since the 1990s, a great deal of efforts have been devoted to the investigation of the GS and GC problems, with many advanced results obtained and many advanced theories developed. Nevertheless, it has been noticed that, about these topics, a fundamental question remains: if two arrays of continuous or discrete dynamical systems can achieve synchronization or consensus, particularly GS or GC, what kind of generic representations should these systems have? In our two recent papers published in the *International Journal of Bifurcation and Chaos*, fairly complete answers were provided, which constitute the main contents of this monograph.

Specifically, this monograph introduces important notions of GS and GC, presented in the following two parts, respectively.

Part 1 first introduces the concepts and mathematical descriptions of GS for coupled discrete array of difference systems (CDADS), coupled continuous array of differential systems (CCADS), non-directional CDADS/ CCADS, bidirectional CDADS/CCADS, non-autonomous CDADS/ CCADS, and non-autonomous bidirectional CDADS/ CCADS. Then, it develops eleven GS theorems on CDADS and CCADS. As applications, seven numerical simulation examples are presented, based on the GS theorems for the corresponding CDADS and CCADS.

Part 2 first introduces the concepts and mathematical descriptions of GC for coupled discrete array of difference systems (CDADS), coupled continuous array of differential systems (CCADS), non-directional CDADS/ CCADS, bidirectional CDADS/CCADS, non-autonomous CDADS/ CCADS, and non-autonomous bidirectional CDADS/ CCADS. Then, it develops eleven GS theorems on CDADS and CCADS. As applications, seven numerical simulation examples are presented, based on the GS theorems for the corresponding CDADS and CCADS. It develops ten GC theorems on CDADS and CCADS. As applications, six numerical simulation examples are presented, based on the GC theorems for the corresponding CDADS and CCADS.

This monograph is suitable for researchers and practitioners undertaking the studies of synchronization and consensus of multi-agent systems, graduate students and senior undergraduate students with the backgrounds in calculus, linear algebra and ordinary differential equations, equipped with computer programming skills, in mathematics, physics, engineering and even social sciences.

Acknowledgments

This book and the associated research were financially supported by the National Natural Science Foundations of China (Grant Nos. 61074192, 61170037) and the Hong Kong Research Grants Council under GRF Grants (CityU 11208515, 11234916, 11200317).

Guanrong Chen
Lequan Min

Abbreviations

CCADS — Coupled Continuous Array of Differential Systems

CDADS — Coupled Discrete Array of Difference Systems

GC — Generalized Consensus

GS — Generalized Synchronization

PGC — Partial Generalized Consensus

PGS — Partial Generalized Synchronization

Contents

Chapter 1

Generalized Synchronization in an Array of Nonlinear Dynamic Systems with Applications to Chaotic CNN

1.1 Introduction

The classic notion of *synchronization*, initiated from the Dutch scientists Christian Huygens in 1665, has been extensively investigated, in various fields of natural and social sciences, engineering and technology. Many fundamental theories and effective methodologies for both state and phase synchronizations have been developed.

Starting from synchronization between two coupled systems, the investigation has gradually evolved to synchronization between two coupled arrays of systems [Wu and Chua (1995)], and to complex networks of systems [Wu (2008)]. This monograph focuses on the study of two arrays of nonlinear dynamical systems, revealing the relationships between the two arrays of systems, which could be either autonomous or non-autonomous, with regard to synchronization and consensus. Conceptually, synchronization means that the dynamical behaviors of two system arrays trend to be identical as time evolves, while consensus means that the difference between the dynamics of the two arrays becomes sufficiently small when the processing time is long enough, where the latter becomes the former if the small difference actually becomes zero.

Regarding the nonlinear dynamical systems, more precisely chaotic systems will be considered in this monograph. Since the pioneering work of Pecora and Carroll [Pecora and Carroll (1990)], chaos synchronization has been extensively studied, in for instance [Yang and Chua (1996); Hunt *et al.* (1997); Chen and Dong (1998); Murali and Laskshmanan (1998); Yang and Chua (1999); Wu and Chua (1993); Li and Chen (2003); Chee and Xu (2006); Gao *et al.* (2006); Ge and Lin (2007); Ji *et al.* (2008)], to name just some representative works. Chaos synchronization has found potential applications in many engineering and technology systems [Chen and

Dong (1998); Cui *et al.* (2016); Shahverdiev (2019)], for example in secure communications [Pecora and Carroll (1990); Yang and Chua (1996); Lau and Tse (2003); Gámez-Guzmán *et al.* (2009); Chen *et al.* (2017)], biological science [Pei *et al.* (1997); Liu and Chen (2003); Sausedo-Solorio and Pisarchik (2014)], laser devices and systems [Imai *et al.* (2003); Wu and Zhu (2003); Gross *et al.* (2006); Li *et al.* (2014)], chemical reactions [Li and Chen (2003); Hu and Cao (2016)], information processing [Wu and Chua (1993); Chen (2005); Na *et al.* (2009)], and so on.

In a more general setting, the generalized synchronization (GS) and generalized consensus (GC) respectively, will be considered, particularly on chaotic systems. Generally, GS or GC is referred to synchronization or consensus by comparing a certain variant or transform of the dynamics of one array to that of the other array, which will be precisely defined below in this part of the book. In retrospect, since the 1990s, great efforts have been devoted to the investigation of both GS and GC problems, especially with chaotic systems, having many advanced results obtained and many advanced theories developed. In this part of the text, the basic concepts and mathematical descriptions of GS will be introduced, for coupled discrete array of difference systems (CDADS), coupled continuous array of differential systems (CCADS), non-directional CDADS and CCADS, bidirectional CDADS and CCADS, non-autonomous CDADS and CCADS, and non-autonomous bidirectional CDADS and CCADS, respectively. Eleven basic GS theorems on CDADS and CCADS will be established, and seven numerical simulation examples will be presented.

The research interest on synchronization in complex networks has also seen rapid increase, especially regarding small-world and scale-free networks [Lago-Fermandez *et al.* (2000); Wang and Chen (2002a,b)].

In the investigations of the above problems, a general question is: if two arrays of systems can achieve GS or GC with respect to a transformation, what kind of representations should these systems have? To answer this question, some constructive GS theorems will be presented in this part of the book, which were developed previously for vector differential equation systems [Zhang and Min (2000)], vector difference discrete systems [Zang *et al.* (2007)], arrays of difference discrete systems [Zang and Min (2008)], bidirectional vector differential continuous systems and difference discrete systems [Ji *et al.* (2008)], arrays of differential continuous systems [Min and Zang (2009)], and bidirectional arrays of differential continuous systems and arrays of difference discrete systems [Zang *et al.* (2012, 2013)], and so on. Some of these results will be summarized and further extended, with applications demonstrated.

The rest of this part is organized as follows. Section 1.2 introduces some definitions on space matrices, GS of CDADS, GS of CCADS, and GS transformations. Section 1.3 introduces four GS theorems for difference and differentiable systems, and derives seven new GS theorems for coupled arrays of difference and differentiable systems. As applications, seven autonomous or non-autonomous GS of CDADS and GS of CCAS are simulated and demonstrated in Section 1.4. Concluding remarks are presented in Section 1.5.

1.2 Basic Concepts and Definitions

To establish some GS theorems for CDADS and CCADS, new concepts and definitions are first introduced [Min and Zang (2009); Zang *et al.* (2012)].

Definition 1.1. Let l be a fixed positive integer and

$$\mathbf{X}^l = (x_{l\,i,j})_{M\times N}$$

be an $M \times N$ matrix. Then,

$$\mathbf{X} = (\mathbf{X}^1, \mathbf{X}^2, \cdots, \mathbf{X}^n)^{\mathrm{T}} \triangleq (x_{l\,i,j})_{n\times M\times N}$$

is called a space matrix, with elements $x_{l\,i,j}$, where $l = 1, 2, ..., n; i = 1, 2, ..., M; j = 1, 2, ..., N$.

Definition 1.2. A space of space matrices $\mathbb{R}^{n\times M\times N}$ is the set of all space matrices $(x_{l\,i,j})_{n\times M\times N}$. Define

$$\mathbf{X} + \mathbf{Y} = (x_{l\,i,j} + y_{l\,i,j})_{n\times M\times N},$$
$$\alpha\mathbf{X} = (\alpha x_{l\,i,j})_{n\times M\times N}, \quad \alpha \in \mathbb{R}.$$

Then, $\mathbb{R}^{n\times M\times N}$ is a real vector space of dimension $n \times M \times N$. This space is equipped with a norm $\|\cdot\|$, where it is noted that all norms on a finite-dimensional space are equivalent.

Definition 1.3. Let $H : \mathbb{R}^{n\times M\times N} \to \mathbb{R}^{n\times M\times N}$ be a transformation (see Fig. 1.2.1). Then, for every $\mathbf{X} \in \mathbb{R}^{n\times M\times N}$, one can write

$$H(\mathbf{X}) = (h_{l\,i,j}(\mathbf{X})),$$

or

$$H(\mathbf{X}) = (h_{l\,i,j}(\mathbf{X}))_{n\times M\times N}.$$

where $l = 1, 2, \cdots, n$; $i = 1, 2, \cdots, M$; $j = 1, 2, \cdots, N$.

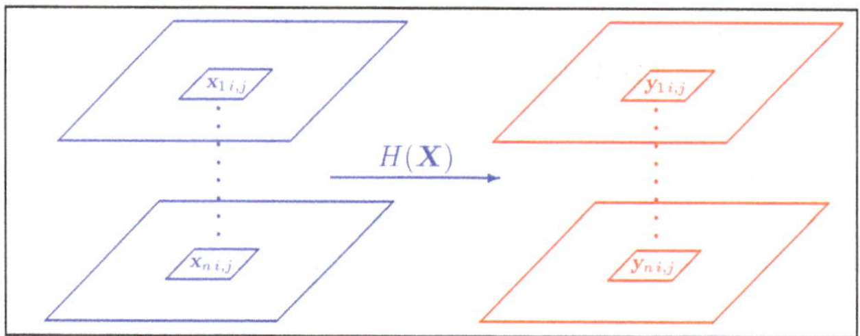

Fig. 1.2.1: Transformation H is a mapping, which maps a space matrix $\mathbf{X} = (x_{l\,i,j})_{n\times M\times N}$ to another space matrix $\mathbf{X} = (y_{l\,i,j})_{n\times M\times N}$.

1.2.1 *GS of a Discrete Array of Difference Systems*

This subsection introduces the concepts of GS of CDADS, bidirectional CDADS, non-autonomously CDADS, and non-autonomously bidirectional CDADS.

First, the GS pf CDADS is studied.

(1) CDADS

Definition 1.4. A CDADS consists of two parts:

$$\begin{cases} x_{1\,i,j}(k+1) = f_{1\,i,j}(\mathbf{X}(k)) \\ x_{2\,i,j}(k+1) = f_{2\,i,j}(\mathbf{X}(k)) \\ \quad\vdots \\ x_{m\,i,j}(k+1) = f_{m\,i,j}(\mathbf{X}(k)) \\ \quad\vdots \\ x_{n\,i,j}(k+1) = f_{n\,i,j}(\mathbf{X}(k)) \end{cases} \tag{1.2.1}$$

and

$$\begin{cases} y_{1\,i,j}(k+1) = g_{1\,i,j}(\mathbf{Y}(k), \mathbf{X}(k)) \\ y_{2\,i,j}(k+1) = g_{2\,i,j}(\mathbf{Y}(k), \mathbf{X}(k)) \\ \quad\vdots \\ y_{m\,i,j}(k+1) = g_{m\,i,j}(\mathbf{Y}(k), \mathbf{X}(k)), \end{cases} \tag{1.2.2}$$

where

$$
\begin{aligned}
\boldsymbol{X}(k) &= (\boldsymbol{X}^1(k), \boldsymbol{X}^2(k), \cdots, \boldsymbol{X}^n(k))^{\mathrm{T}} \\
&= (x_{l\,i,j}(k))_{n\times M\times N}, \qquad (1.2.3)\\
\boldsymbol{Y}(k) &= (\boldsymbol{Y}^1(k), \boldsymbol{Y}^2(k), \cdots, \boldsymbol{Y}^m(k))^{\mathrm{T}} \\
&= (y_{l\,i,j}(k))_{m\times M\times N}, \qquad (1.2.4)\\
& m \le n, \quad i = 1, 2, \ldots, M, \quad j = 1, 2, \ldots, N.
\end{aligned}
$$

Here, $\boldsymbol{X}$ and $\boldsymbol{Y}$ are space matrices. In compact forms, CDADS (1.2.1) and (1.2.2) can be written as

$$\boldsymbol{X}(k+1) = F(\boldsymbol{X}(k)), \qquad (1.2.5)$$

$$\boldsymbol{Y}(k+1) = G(\boldsymbol{Y}(k), \boldsymbol{X}(k)), \qquad (1.2.6)$$

where

$$F(\boldsymbol{X}(k)) = ((f_{1\,i,j}(\boldsymbol{X}(k))_{M\times N}, \ldots, (f_{n\,i,j}(\boldsymbol{X}(k))_{M\times N})^{\mathrm{T}}, \qquad (1.2.7)$$

$$
\begin{aligned}
G(\boldsymbol{Y}(k), \boldsymbol{X}(k)) = ((g_{1\,i,j}(\boldsymbol{Y}(k), (\boldsymbol{X}(k))_{M\times N}, \ldots,\\
(g_{m\,i,j}(\boldsymbol{Y}(k), \boldsymbol{X}(k)))_{M\times N})^{\mathrm{T}}. \qquad (1.2.8)
\end{aligned}
$$

Definition 1.5. The CDADS defined by (1.2.5) and (1.2.6) are said to be in GS with respect to a transformation $H : \mathbb{R}^{m\times M\times N} \to \mathbb{R}^{m\times M\times N}$ (see Fig. 1.2.2), if there exists an open subset $B \subset \mathbb{R}^{n\times M\times N} \times \mathbb{R}^{m\times M\times N}$ such that, for any trajectory $(\boldsymbol{X}(k), \boldsymbol{Y}(k))$ of systems (1.2.5) and (1.2.6) with initial condition $(\boldsymbol{X}(0), \boldsymbol{Y}(0)) \in B$, one has

$$\lim_{k\to+\infty} \|H(\boldsymbol{X}_m(k)) - \boldsymbol{Y}(k)\| = 0, \qquad (1.2.9)$$

where

$$
\begin{aligned}
\boldsymbol{X}_m(k) &= (\boldsymbol{X}^1(k), \cdots, \boldsymbol{X}^m(k))^{\mathrm{T}} \\
&= (x_{l\,i,j}(k))_{m\times M\times N}, \\
\boldsymbol{Y}(k) &= (\boldsymbol{Y}^1(k), \cdots, \boldsymbol{Y}^m(k))^{\mathrm{T}} \\
&= (y_{l\,i,j}(k))_{m\times M\times N}.
\end{aligned}
$$

Remark 1.2.1. Condition (1.2.9) implies that, if the initial condition $\boldsymbol{Y}(0) = H(\boldsymbol{X}_m(0))$, then for any $k \ge 1, H(\boldsymbol{X}_m(k)) = \boldsymbol{Y}(k)$.

(2) Bidirectional CDADS

Definition 1.6. A bidirectional CDADS consists of two parts:

$$\begin{cases} x_{1\,i,j}(k+1) = f_{1\,i,j}(\boldsymbol{X}(k), \boldsymbol{Y}(k)) \\ x_{2\,i,j}(k+1) = f_{2\,i,j}(\boldsymbol{X}(k), \boldsymbol{Y}(k)) \\ \quad\vdots \\ x_{m\,i,j}(k+1) = f_{m\,i,j}(\boldsymbol{X}(k), \boldsymbol{Y}(k)) \\ \quad\vdots \\ x_{n\,i,j}(k+1) = f_{n\,i,j}(\boldsymbol{X}(k), \boldsymbol{Y}(k)) \end{cases} \tag{1.2.10}$$

and

$$\begin{cases} y_{1\,i,j}(k+1) = g_{1\,i,j}(\boldsymbol{Y}(k), \boldsymbol{X}(k)) \\ y_{2\,i,j}(k+1) = g_{2\,i,j}(\boldsymbol{Y}(k), \boldsymbol{X}(k)) \\ \quad\vdots \\ y_{m\,i,j}(k+1) = g_{m\,i,j}(\boldsymbol{Y}(k), \boldsymbol{X}(k)), \end{cases} \tag{1.2.11}$$

where

$$\begin{aligned} \boldsymbol{X}(k) &= (\boldsymbol{X}^1(k), \boldsymbol{X}^2(k), \cdots, \boldsymbol{X}^n(k))^{\mathrm{T}} \\ &= (x_{l\,i,j}(k))_{n\times M\times N}, && (1.2.12) \\ \boldsymbol{Y}(k) &= (\boldsymbol{Y}^1(k), \boldsymbol{Y}^2(k), \cdots, \boldsymbol{Y}^m(k))^{\mathrm{T}} \\ &= (y_{l\,i,j}(k))_{m\times M\times N}, && (1.2.13) \\ & \quad m \le n,\ i = 1, 2, \ldots, M,\ \ j = 1, 2, \ldots, N. \end{aligned}$$

In compact forms, the bidirectional CDADS (1.2.10) and (1.2.11) can be written as

$$\boldsymbol{X}(k+1) = F(\boldsymbol{X}(k), \boldsymbol{Y}(k)), \tag{1.2.14}$$

$$\boldsymbol{Y}(k+1) = G(\boldsymbol{Y}(k), \boldsymbol{X}(k)), \tag{1.2.15}$$

where

$$\begin{aligned} F(\boldsymbol{X}(k), \boldsymbol{Y}(k)) = ((&f_{1\,i,j}(\boldsymbol{X}(k), \boldsymbol{Y}(k)))_{M\times N}, \ldots, \\ &(f_{n\,i,j}(\boldsymbol{X}(k), \boldsymbol{Y}(k))_{M\times N})^{\mathrm{T}} && (1.2.16) \\ G(\boldsymbol{Y}(k), \boldsymbol{X}(k)) = ((&g_{1\,i,j}(\boldsymbol{Y}(k), \boldsymbol{X}(k))_{M\times N}, \ldots, \\ &(g_{m\,i,j}(\boldsymbol{Y}(k), \boldsymbol{X}(k)))_{M\times N})^{\mathrm{T}}. && (1.2.17) \end{aligned}$$

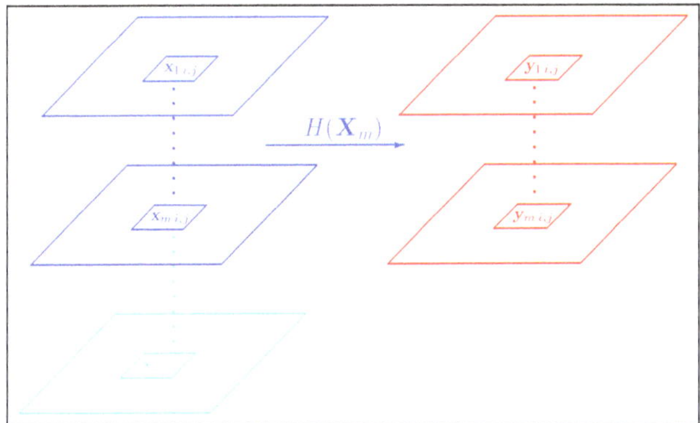

Fig. 1.2.2: Transformation H is a mapping, which maps a subset B in the space of space $\mathbb{R}^{m\times M\times N}$ into the space of space matrices $\mathbb{R}^{m\times M\times N}$.

Definition 1.7. The bidirectional CDADS defined by (1.2.14) and (1.2.15) are said to be in GS with respect to a transformation $H:\mathbb{R}^{m\times M\times N}\to\mathbb{R}^{m\times M\times N}$ (see Fig. 1.2.2), if there exists an open subset $B\subset\mathbb{R}^{n\times M\times N}\times\mathbb{R}^{m\times M\times N}$ such that, for any trajectory $(\boldsymbol{X}(k),\boldsymbol{Y}(k))$ of systems (1.2.14) and (1.2.15) with initial condition $(\boldsymbol{X}(0),\boldsymbol{Y}(0))\in B$, one has

$$\lim_{k\to+\infty}\|H(\mathbf{X}_m(k))-\mathbf{Y}(k)\|=0, \tag{1.2.18}$$

where

$$\begin{aligned}\mathbf{X}_m(k)&=(\mathbf{X}^1(k),\cdots,\mathbf{X}^m(k))^{\mathrm{T}}\\&=(x_{l\,i,j}(k))_{m\times M\times N},\\ \mathbf{Y}(k)&=(\mathbf{Y}^1(k),\cdots,\mathbf{Y}^m(k))^{\mathrm{T}}\\&=(y_{l\,i,j}(k))_{m\times M\times N}.\end{aligned}$$

(3) Non-autonomous CDADS

Definition 1.8. A non-autonomously CDADS consists of two parts:

$$\begin{cases}x_{1\,i,j}(k+1)=f_{1\,i,j}(\mathbf{X}(k),k)\\ x_{2\,i,j}(k+1)=f_{2\,i,j}(\mathbf{X}(k),k)\\ \qquad\vdots\\ x_{m\,i,j}(k+1)=f_{m\,i,j}(\mathbf{X}(k),k)\\ \qquad\vdots\\ x_{n\,i,j}(k+1)=f_{n\,i,j}(\mathbf{X}(k),k)\end{cases} \tag{1.2.19}$$

and

$$\begin{cases} y_{1\,i,j}(k+1) = g_{1\,i,j}(\mathbf{Y}(k), \mathbf{X}(k), k) \\ y_{2\,i,j}(k+1) = g_{2\,i,j}(\mathbf{Y}(k), \mathbf{X}(k), k) \\ \quad\vdots \\ y_{m\,i,j}(k+1) = g_{m\,i,j}(\mathbf{Y}(k), \mathbf{X}(k), k), \end{cases} \tag{1.2.20}$$

where

$$\begin{aligned} \mathbf{X}(k) &= (\mathbf{X}^1(k), \mathbf{X}^2(k), \cdots, \mathbf{X}^n(k))^{\mathrm{T}} \\ &= (x_{l\,i,j}(k))_{n\times M\times N}, & (1.2.21) \\ \mathbf{Y}(k) &= (\mathbf{Y}^1(k), \mathbf{Y}^2(k), \cdots, \mathbf{Y}^m(k))^{\mathrm{T}} \\ &= (y_{l\,i,j}(k))_{m\times M\times N}, & (1.2.22) \\ & \quad m \le n,\ i = 1, 2, \ldots, M,\ j = 1, 2, \ldots, N. \end{aligned}$$

In compact forms, the non-autonomous CDADS (1.2.19) and (1.2.20) can be written as

$$\mathbf{X}(k+1) = F(\mathbf{X}(k), k), \tag{1.2.23}$$

$$\mathbf{Y}(k+1) = G(\mathbf{Y}(k), \mathbf{X}(k), k), \tag{1.2.24}$$

where

$$\begin{aligned} F(\mathbf{X}(k), k) = ((f_{1\,i,j}(\mathbf{X}(k), k))_{M\times N}, \ldots, \\ (f_{n\,i,j}(\mathbf{X}(k), k))_{M\times N})^{\mathrm{T}}, & \quad (1.2.25) \\ G(\mathbf{Y}(k), \mathbf{X}(k), k) = ((g_{1\,i,j}(\mathbf{Y}(k), \mathbf{X}(k), k))_{M\times N}, \ldots, \\ (g_{m\,i,j}(\mathbf{Y}(k), \mathbf{X}(k), k))_{M\times N})^{\mathrm{T}}. & \quad (1.2.26) \end{aligned}$$

Definition 1.9. The non-autonomous CDADS defined by (1.2.23) and (1.2.24) are said to be in GS with respect to a transformation $H : \mathbb{R}^{m\times M\times N} \times \mathbb{Z}^+ \to \mathbb{R}^{m\times M\times N}$ (see Fig. 1.2.3), if there exists an open subset $B \subset \mathbb{R}^{n\times M\times N} \times \mathbb{R}^{m\times M\times N}$ such that for any trajectory $(\mathbf{X}(k), \mathbf{Y}(k))$ of systems (1.2.23) and (1.2.24) with initial condition $(\mathbf{X}(0), \mathbf{Y}(0)) \in B$, one has

$$\lim_{k\to+\infty} \|H(\mathbf{X}_m(k), k) - \mathbf{Y}(k)\| = 0, \tag{1.2.27}$$

where

$$\begin{aligned} \mathbf{X}_m(k) &= (\mathbf{X}^1(k), \cdots, \mathbf{X}^m(k))^{\mathrm{T}} \\ &= (x_{l\,i,j}(k))_{m\times M\times N}, \\ \mathbf{Y}(k) &= (\mathbf{Y}^1(k), \cdots, \mathbf{Y}^m(k))^{\mathrm{T}} \\ &= (y_{l\,i,j}(k))_{m\times M\times N}. \end{aligned}$$

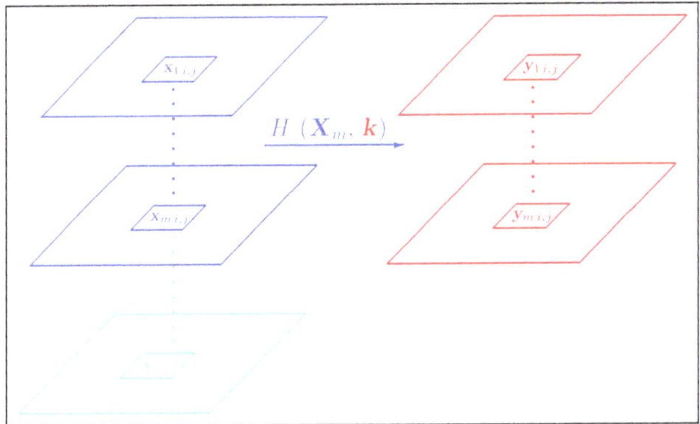

Fig. 1.2.3: Transformation H is a mapping, which maps a subset in the space of space matrices $\mathbb{R}^{m\times M\times N}\times\mathbb{Z}^+$ into the space of space matrices $\mathbb{R}^{m\times M\times N}$.

Remark 1.2.2. As a special case, $H:\mathbb{R}^{m\times M\times N}\to\mathbb{R}^{m\times M\times N}$ may be independent of the iteration step $k\in\mathbb{Z}^+$.

(4) Non-autonomous bidirectional CDADS

Definition 1.10. A non-autonomous bidirectional CDADS consists of two parts:

$$\begin{cases} x_{1\,i,j}(k+1) = f_{1\,i,j}(\mathbf{X}(k),\mathbf{Y}(k),k) \\ x_{2\,i,j}(k+1) = f_{2\,i,j}(\mathbf{X}(k),\mathbf{Y}(k),k) \\ \quad\vdots \\ x_{m\,i,j}(k+1) = f_{m\,i,j}(\mathbf{X}(k),\mathbf{Y}(k),k) \\ \quad\vdots \\ x_{n\,i,j}(k+1) = f_{n\,i,j}(\mathbf{X}(k),\mathbf{Y}(k),k) \end{cases} \tag{1.2.28}$$

and

$$\begin{cases} y_{1\,i,j}(k+1) = g_{1\,i,j}(\mathbf{Y}(k),\mathbf{X}(k),k) \\ y_{2\,i,j}(k+1) = g_{2\,i,j}(\mathbf{Y}(k),\mathbf{X}(k),k) \\ \quad\vdots \\ y_{m\,i,j}(k+1) = g_{m\,i,j}(\mathbf{Y}(k),\mathbf{X}(k),k), \end{cases} \tag{1.2.29}$$

where

$$\begin{aligned}\mathbf{X}(k) &= (\mathbf{X}^1(k),\mathbf{X}^2(k),\cdots,\mathbf{X}^n(k))^{\mathrm{T}} \\ &= (x_{l\,i,j}(k))_{n\times M\times N},\end{aligned} \tag{1.2.30}$$

$$\begin{aligned}\mathbf{Y}(k) &= (\mathbf{Y}^1(k), \mathbf{Y}^2(k), \cdots, \mathbf{Y}^m(k))^{\mathrm{T}} \\ &= (y_{l\,i,j}(k))_{m\times M\times N}, \\ &\quad m \le n,\ i = 1,2,\ldots,M,\ \ j = 1,2,\ldots,N.\end{aligned} \tag{1.2.31}$$

In compact forms, the non-autonomous bidirectional CDADS (1.2.28) and (1.2.29) can be written as

$$\mathbf{X}(k+1) = F(\mathbf{X}(k), \mathbf{Y}(k), k), \tag{1.2.32}$$

$$\mathbf{Y}(k+1) = G(\mathbf{Y}(k), \mathbf{X}(k), k), \tag{1.2.33}$$

where

$$\begin{aligned}F(\mathbf{X}(k), \mathbf{Y}(k), k) = ((&f_{1\,i,j}(\mathbf{X}(k), \mathbf{Y}(k), k))_{M\times N}, \ldots, \\ &(f_{n\,i,j}(\mathbf{X}(k), \mathbf{Y}(k), k))_{M\times N})^{\mathrm{T}},\end{aligned} \tag{1.2.34}$$

$$\begin{aligned}G(\mathbf{Y}(k), \mathbf{X}(k), k) = ((&g_{1\,i,j}(\mathbf{Y}(k), \mathbf{X}(k), k)_{M\times N}, \ldots, \\ &(g_{m\,i,j}(\mathbf{Y}(k), \mathbf{X}(k), k))_{M\times N})^{\mathrm{T}}.\end{aligned} \tag{1.2.35}$$

Definition 1.11. The non-autonomous bidirectional CDADS defined by (1.2.32) and (1.2.33) are said to be in GS with respect to a transformation $H : \mathbb{R}^{m\times M\times N} \times \mathbb{Z}^+ \to \mathbb{R}^{m\times M\times N}$ (see Fig. 1.2.3), if there exists an open subset $B \subset \mathbb{R}^{n\times M\times N} \times \mathbb{R}^{m\times M\times N}$ such that, for any trajectory $(\mathbf{X}(k), \mathbf{Y}(k))$ of systems (1.2.32) and (1.2.33) with initial condition $(\mathbf{X}(0), \mathbf{Y}(0)) \in B$, one has

$$\lim_{k\to+\infty} \|H(\mathbf{X}_m(k), k) - \mathbf{Y}(k)\| = 0, \tag{1.2.36}$$

where

$$\begin{aligned}\mathbf{X}_m(k) &= (\mathbf{X}^1(k), \cdots, \mathbf{X}^m(k))^{\mathrm{T}} \\ &= (x_{l\,i,j}(k))_{m\times M\times N}, \\ \mathbf{Y}(k) &= (\mathbf{Y}^1(k), \cdots, \mathbf{Y}^m(k))^{\mathrm{T}} \\ &= (y_{l\,i,j}(k))_{m\times M\times N}.\end{aligned}$$

Remark 1.2.3. As a special case, $H : \mathbb{R}^{m\times M\times N} \to \mathbb{R}^{m\times M\times N}$ may be independent of the iteration step $k \in \mathbb{Z}^+$.

1.2.2 *GS of a Continuous Array of Differential Systems*

This subsection introduces the notion of GS of CCADS, bidirectional CCADS, non-autonomous CCADS, and non-autonomously bidirectional CCADS.

First, the CCADS is studied.

(1) CCADS

Definition 1.12. A CCADS consists of two parts:

$$\begin{cases} \dot{x}_{1\,i,j}(t) = f_{1\,i,j}(\boldsymbol{X}(t)) \\ \dot{x}_{2\,i,j}(t) = f_{2\,i,j}(\boldsymbol{X}(t)) \\ \quad\vdots \\ \dot{x}_{m\,i,j}(t) = f_{m\,i,j}(\boldsymbol{X}(t)) \\ \quad\vdots \\ \dot{x}_{n\,i,j}(t) = f_{n\,i,j}(\boldsymbol{X}(t)) \end{cases} \tag{1.2.37}$$

and

$$\begin{cases} \dot{y}_{1\,i,j}(t) = g_{1\,i,j}(\boldsymbol{Y}(t),\boldsymbol{X}(t)) \\ \dot{y}_{2\,i,j}(t) = g_{2\,i,j}(\boldsymbol{Y}(t),\boldsymbol{X}(t)) \\ \quad\vdots \\ \dot{y}_{m\,i,j}(t) = g_{m\,i,j}(\boldsymbol{Y}(t),\boldsymbol{X}(t)), \end{cases} \tag{1.2.38}$$

where

$$\begin{aligned} \boldsymbol{X}(t) &= (\boldsymbol{X}^1(t),\boldsymbol{X}^2(t),\cdots,\boldsymbol{X}^n(t))^{\mathrm{T}} \\ &= (x_{l\,i,j}(t))_{n\times M\times N}, \qquad (1.2.39) \\ \boldsymbol{Y}(t) &= (\boldsymbol{Y}^1(t),\boldsymbol{Y}^2(t),\cdots,\boldsymbol{Y}^m(t))^{\mathrm{T}} \\ &= (y_{l\,i,j}(t))_{m\times M\times N}, \qquad (1.2.40) \\ m \le n\,,\ & i=1,2,\ldots,M,\ j=1,2,\ldots,N. \end{aligned}$$

Here, $\boldsymbol{X}(t)$ and $\boldsymbol{Y}(t)$ are space matrices.

In compact forms, CCADS (1.2.37) and (1.2.38) can be written as

$$\dot{\boldsymbol{X}}(t) = F(\boldsymbol{X}(t)), \tag{1.2.41}$$

$$\dot{\boldsymbol{Y}}(t) = G(\boldsymbol{Y}(t),\boldsymbol{X}(t)), \tag{1.2.42}$$

where

$$F(\boldsymbol{X}(t)) = ((f_{1\,i,j}(\boldsymbol{X})(t))_{M\times N},\ldots,(f_{n\,i,j}(\boldsymbol{X})(t))_{M\times N})^{\mathrm{T}}, \tag{1.2.43}$$

$$\begin{aligned} G(\boldsymbol{Y}(t),\boldsymbol{X}(t)) = ((&g_{1\,i,j}(\boldsymbol{Y}(t),\boldsymbol{X}(t))_{M\times N}, \\ &\ldots,(g_{m\,i,j}(\boldsymbol{Y}(t),\boldsymbol{X}(t))_{M\times N})^{\mathrm{T}}. \end{aligned} \tag{1.2.44}$$

Definition 1.13. The CCADS defined by (1.2.41) and (1.2.42) are said to be in GS with respect to a transformation $H:\mathbb{R}^{m\times M\times N}\to\mathbb{R}^{m\times M\times N}$ (see Fig. 1.2.2), if there exists an open subset $B\subset\mathbb{R}^{n\times M\times N}\times\mathbb{R}^{m\times M\times N}$ such that, for any trajectory $(\mathbf{X}(k), \mathbf{Y}(k))$ of systems (1.2.41) and (1.2.42) with initial condition $(\mathbf{X}(0), \mathbf{Y}(0)) \in B$, one has

$$\lim_{t\to+\infty} \|H(\mathbf{X}_m(t)) - \mathbf{Y}(t)\| = 0, \tag{1.2.45}$$

where

$$\mathbf{X}_m(t) = (x_{l\,i,j}(t))_{m\times M\times N},$$
$$\mathbf{Y}(t) = (y_{l\,i,j}(t))_{m\times M\times N}.$$

(2) Bidirectional CCADS

Definition 1.14. A bidirectional CCADS consists of two parts:

$$\begin{cases} \dot{x}_{1\,i,j}(t) = f_{1\,i,j}(\mathbf{X}(t), \mathbf{Y}(t)) \\ \dot{x}_{2\,i,j}(t) = f_{2\,i,j}(\mathbf{X}(t), \mathbf{Y}(t)) \\ \quad\vdots \\ \dot{x}_{m\,i,j}(t) = f_{m\,i,j}(\mathbf{X}(t), \mathbf{Y}(t)) \\ \quad\vdots \\ \dot{x}_{n\,i,j}(t) = f_{n\,i,j}(\mathbf{X}(t), \mathbf{Y}(t)) \end{cases} \tag{1.2.46}$$

and

$$\begin{cases} \dot{y}_{1\,i,j}(t) = g_{1\,i,j}(\mathbf{Y}(t), \mathbf{X}(t)) \\ \dot{y}_{2\,i,j}(t) = g_{2\,i,j}(\mathbf{Y}(t), \mathbf{X}(t)) \\ \quad\vdots \\ \dot{y}_{m\,i,j}(t) = g_{m\,i,j}(\mathbf{Y}(t), \mathbf{X}(t)), \end{cases} \tag{1.2.47}$$

where

$$\begin{aligned} \mathbf{X}(t) &= (\mathbf{X}^1(t), \mathbf{X}^2(t), \cdots, \mathbf{X}^n(t))^{\mathrm{T}} \\ &= (x_{l\,i,j}(t))_{n\times M\times N}, \qquad (1.2.48) \\ \mathbf{Y}(t) &= (\mathbf{Y}^1(t), \mathbf{Y}^2(t), \cdots, \mathbf{Y}^m(t))^{\mathrm{T}} \\ &= (y_{l\,i,j}(t))_{m\times M\times N}, \qquad (1.2.49) \\ m &\le n,\ i = 1, 2, \ldots, M,\ \ j = 1, 2, \ldots, N. \end{aligned}$$

In compact forms, the bidirectional CCADS (1.2.46) and (1.2.47) can be written as

$$\dot{\mathbf{X}}(t) = F(\mathbf{X}(t), \mathbf{Y}(t)), \tag{1.2.50}$$
$$\dot{\mathbf{Y}}(t) = G(\mathbf{Y}(t), \mathbf{X}(t)), \tag{1.2.51}$$

where

$$\begin{aligned} F(\mathbf{X}(t), \mathbf{Y}(t)) = ((&f_{1\,i,j}(\mathbf{X}(t), \mathbf{Y}(t)))_{M\times N}, \ldots, \\ &(f_{n\,i,j}(\mathbf{X}(t), \mathbf{Y}(t)))_{M\times N})^{\mathrm{T}}, \end{aligned} \tag{1.2.52}$$

$$\begin{aligned} G(\mathbf{Y}(t), \mathbf{X}(t)) = ((&g_{1\,i,j}(\mathbf{Y}(t), \mathbf{X}(t)))_{M\times N}, \ldots, \\ &(g_{m\,i,j}(\mathbf{Y}(t), \mathbf{X}(t))_{M\times N})^{\mathrm{T}}. \end{aligned} \tag{1.2.53}$$

Definition 1.15. The bidirectional CCADS defined by (1.2.50) and (1.2.51) are said to be in GS with respect to a transformation $H : \mathbb{R}^{m\times M\times N} \to \mathbb{R}^{m\times M\times N}$ (see Fig. 1.2.2), if there exists an open subset $B \subset \mathbb{R}^{n\times M\times N} \times \mathbb{R}^{m\times M\times N}$ such that, for any trajectory $(\mathbf{X}(k), \mathbf{Y}(k))$ of systems (1.2.50) and (1.2.51) with initial condition $(\mathbf{X}(0), \mathbf{Y}(0)) \in B$, one has

$$\lim_{t\to+\infty} \|H(\mathbf{X}_m(t)) - \mathbf{Y}(t)\| = 0, \tag{1.2.54}$$

where

$$\begin{aligned} \mathbf{X}_m(t) &= (x_{l\,i,j}(t))_{m\times M\times N}, \\ \mathbf{Y}(t) &= (y_{l\,i,j}(t))_{m\times M\times N}. \end{aligned}$$

(3) Non-autonomous CCADS

Definition 1.16. A non-autonomous CCADS consists of two parts:

$$\begin{cases} \dot{x}_{1\,i,j}(t) = f_{1\,i,j}(\mathbf{X}(t), t) \\ \dot{x}_{2\,i,j}(t) = f_{2\,i,j}(\mathbf{X}(t), t) \\ \quad\vdots \\ \dot{x}_{m\,i,j}(t) = f_{m\,i,j}(\mathbf{X}(t), t) \\ \quad\vdots \\ \dot{x}_{n\,i,j}(t) = f_{n\,i,j}(\mathbf{X}(t), t) \end{cases} \tag{1.2.55}$$

and

$$\begin{cases} \dot{y}_{1\,i,j}(t) = g_{1\,i,j}(\mathbf{Y}(t), \mathbf{X}(t), t) \\ \dot{y}_{2\,i,j}(t) = g_{2\,i,j}(\mathbf{Y}(t), \mathbf{X}(t), t) \\ \quad\vdots \\ \dot{y}_{m\,i,j}(t) = g_{m\,i,j}(\mathbf{Y}(t), \mathbf{X}(t), t), \end{cases} \tag{1.2.56}$$

where

$$\begin{aligned} \mathbf{X}(t) &= (\mathbf{X}^1(t), \mathbf{X}^2(t), \cdots, \mathbf{X}^n(t))^{\mathrm{T}} \\ &= (x_{l\,i,j}(t))_{n\times M\times N}, \end{aligned} \tag{1.2.57}$$

$$\begin{aligned} \mathbf{Y}(t) &= (\mathbf{Y}^1(t), \mathbf{Y}^2(t), \cdots, \mathbf{Y}^m(t))^{\mathrm{T}} \\ &= (y_{l\,i,j}(t))_{m\times M\times N}, \end{aligned} \tag{1.2.58}$$

$$m \le n,\ i = 1, 2, \ldots, M,\ \ j = 1, 2, \ldots, N.$$

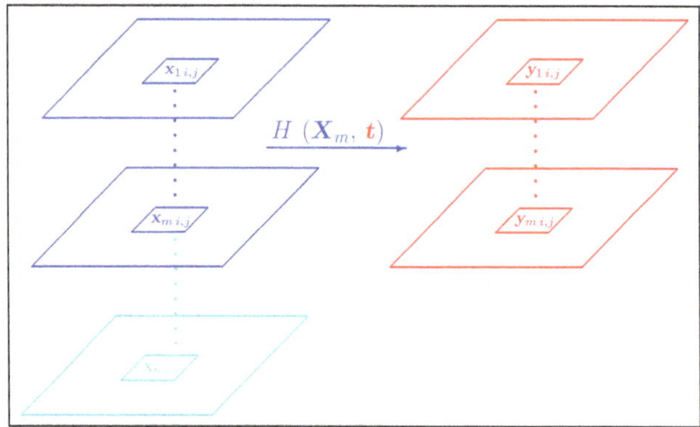

Fig. 1.2.4: Transformation H is a mapping, which maps a subset in the space of space matrices $\mathbb{R}^{m\times M\times N}\times\mathbb{R}^+$ into the space of space matrices $\mathbb{R}^{m\times M\times N}$.

In compact forms, the non-autonomously CCADS (1.2.55) and (1.2.56) can be written as

$$\dot{\mathbf{X}}(t) = F(\mathbf{X}(t), t), \tag{1.2.59}$$

$$\dot{\mathbf{Y}}(t) = G(\mathbf{Y}(t), \mathbf{X}(t), t), \tag{1.2.60}$$

where

$$F(\mathbf{X}(t), t) = ((f_{1\,i,j}(\mathbf{X})(t), t))_{M\times N}, \ldots, (f_{n\,i,j}(\mathbf{X})(t), t))_{M\times N})^{\mathrm{T}}, \tag{1.2.61}$$

$$G(\mathbf{Y}(t), \mathbf{X}(t), t) = ((g_{1\,i,j}(\mathbf{Y}(t), \mathbf{X}(t), t))_{M\times N}, \ldots, (g_{m\,i,j}(\mathbf{Y}(t), \mathbf{X}(t), t))_{M\times N})^{\mathrm{T}}. \tag{1.2.62}$$

Definition 1.17. A non-autonomous CCADS defined by (1.2.59) and (1.2.60) are said to be in GS with respect to a transformation $H : \mathbb{R}^{m\times M\times N}\times\mathbb{R}^+\to\mathbb{R}^{m\times M\times N}$ (see Fig. 1.2.4), if there exists an open subset $B\subset\mathbb{R}^{n\times M\times N}\times\mathbb{R}^{m\times M\times N}$ such that, for any trajectory $(\mathbf{X}(k), \mathbf{Y}(k))$ of systems (1.2.59) and (1.2.60) with initial condition $(\mathbf{X}(0), \mathbf{Y}(0))\in B$, one has

$$\lim_{t\to+\infty}\|H(\mathbf{X}_m(t), t) - \mathbf{Y}(t)\| = 0, \tag{1.2.63}$$

where

$$\mathbf{X}_m(t) = (x_{l\,i,j}(t))_{m\times M\times N},$$
$$\mathbf{Y}(t) = (y_{l\,i,j}(t))_{m\times M\times N}.$$

Remark 1.2.4. As a special case, the transformation $H : \mathbb{R}^{m\times M\times N} \to \mathbb{R}^{m\times M\times N}$ may be independent of the time variable $t \in [0, \infty]$.

(4) Non-autonomous Bidirectional CCADS

Definition 1.18. A non-autonomous bidirectional CCADS consists of two parts:

$$\begin{cases} \dot{x}_{1\,i,j}(t) &= f_{1\,i,j}(\boldsymbol{X}(t), \boldsymbol{Y}(t), t) \\ \dot{x}_{2\,i,j}(t) &= f_{2\,i,j}(\boldsymbol{X}(t), \boldsymbol{Y}(t), t) \\ \quad\vdots & \\ \dot{x}_{m\,i,j}(t) &= f_{m\,i,j}(\boldsymbol{X}(t), \boldsymbol{Y}(t), t) \\ \quad\vdots & \\ \dot{x}_{n\,i,j}(t) &= f_{n\,i,j}(\boldsymbol{X}(t), \boldsymbol{Y}(t), t) \end{cases} \tag{1.2.64}$$

and

$$\begin{cases} \dot{y}_{1\,i,j}(t) &= g_{1\,i,j}(\boldsymbol{Y}(t), \boldsymbol{X}(t), t) \\ \dot{y}_{2\,i,j}(t) &= g_{2\,i,j}(\boldsymbol{Y}(t), \boldsymbol{X}(t), t) \\ \quad\vdots & \\ \dot{y}_{m\,i,j}(t) &= g_{m\,i,j}(\boldsymbol{Y}(t), \boldsymbol{X}(t), t), \end{cases} \tag{1.2.65}$$

where

$$\begin{aligned} \boldsymbol{X}(t) &= (\boldsymbol{X}^1(t), \boldsymbol{X}^2(t), \cdots, \boldsymbol{X}^n(t))^{\mathrm{T}} \\ &= (x_{l\,i,j}(t))_{n\times M\times N}, \qquad (1.2.66) \\ \boldsymbol{Y}(t) &= (\boldsymbol{Y}^1(t), \boldsymbol{Y}^2(t), \cdots, \boldsymbol{Y}^m(t))^{\mathrm{T}} \\ &= (y_{l\,i,j}(t))_{m\times M\times N}, \qquad (1.2.67) \\ m &\le n,\ i = 1, 2, \ldots, M,\ \ j = 1, 2, \ldots, N. \end{aligned}$$

In compact forms, the non-autonomous bidirectional CCADS (1.2.64) and (1.2.65) can be written as

$$\dot{\boldsymbol{X}}(t) = F(\boldsymbol{X}(t), \boldsymbol{Y}(t), t), \tag{1.2.68}$$

$$\dot{\boldsymbol{Y}}(t) = G(\boldsymbol{Y}(t), \boldsymbol{X}(t), t), \tag{1.2.69}$$

where

$$\begin{aligned} F(\boldsymbol{X}(t), \boldsymbol{Y}(t), t) &= ((f_{1\,i,j}(\boldsymbol{X}(t), \boldsymbol{Y}(t), t))_{M\times N}, \ldots, \\ &\quad (f_{n\,i,j}(\boldsymbol{X}(t), \boldsymbol{Y}(t), t))_{M\times N})^{\mathrm{T}}, \qquad (1.2.70) \\ G(\boldsymbol{Y}(t), \boldsymbol{X}(t)) &= ((g_{1\,i,j}(\boldsymbol{Y}(t), \boldsymbol{X}(t), t))_{M\times N}, \ldots, \\ &\quad (g_{m\,i,j}(\boldsymbol{Y}(t), \boldsymbol{X}(t), t))_{M\times N})^{\mathrm{T}}. \qquad (1.2.71) \end{aligned}$$

Definition 1.19. The non-autonomous bidirectional CCADS defined by (1.2.68) and (1.2.69) are said to be in GS with respect to a transformation $H : \mathbb{R}^{m\times M\times N} \times \mathbb{R}^{+} \to \mathbb{R}^{m\times M\times N}$ (see Fig. 1.2.4), if there exists an open subset $B \subset \mathbb{R}^{n\times M\times N} \times \mathbb{R}^{m\times M\times N}$ such that, for any trajectory $(\mathbf{X}(k), \mathbf{Y}(k))$ of systems (1.2.68) and (1.2.69) with initial condition $(\mathbf{X}(0), \mathbf{Y}(0)) \in B$, one has

$$\lim_{t\to+\infty} \|H(\mathbf{X}_m(t), t) - \mathbf{Y}(t)\| = 0, \tag{1.2.72}$$

where

$$\begin{aligned} \mathbf{X}_m(t) &= (x_{l\,i,j}(t))_{m\times M\times N}, \\ \mathbf{Y}(t) &= (y_{l\,i,j}(t))_{m\times M\times N}. \end{aligned}$$

Remark 1.2.5. As a special case, the transformation $H : \mathbb{R}^{m\times M\times N} \to \mathbb{R}^{m\times M\times N}$ may be independent of the time variable $t \in [0, \infty]$.

1.2.3 *PGS of Non-autonomous Bidirectional CDADS*

Definition 1.20. A general non-autonomous bidirectional coupled discrete array of dynamic systems consists of two parts:

$$\begin{cases} x_{1\,i,j}(k+1) &= f_{1\,i,j}(\mathbf{X}(k), \mathbf{Y}(k), k) \\ x_{2\,i,j}(k+1) &= f_{2\,i,j}(\mathbf{X}(k), \mathbf{Y}(k), k) \\ \quad\vdots & \\ x_{m\,i,j}(k+1) &= f_{m\,i,j}(\mathbf{X}(k), \mathbf{Y}(k), k) \\ \quad\vdots & \\ x_{N_1\,i,j}(k+1) &= f_{N_1\,i,j}(\mathbf{X}(k), \mathbf{Y}(k), k) \end{cases} \tag{1.2.73}$$

and

$$\begin{cases} y_{1\,i,j}(k+1) &= g_{1\,i,j}(\mathbf{Y}(k), \mathbf{X}(k), k) \\ y_{2\,i,j}(k+1) &= g_{2\,i,j}(\mathbf{Y}(k), \mathbf{X}(k), k) \\ \quad\vdots & \\ y_{m\,i,j}(k+1) &= g_{m\,i,j}(\mathbf{Y}(k), \mathbf{X}(k), k) \\ \quad\vdots & \\ y_{N_2\,i,j}(k+1) &= g_{N_2\,i,j}(\mathbf{X}(k), \mathbf{Y}(k), k), \end{cases} \tag{1.2.74}$$

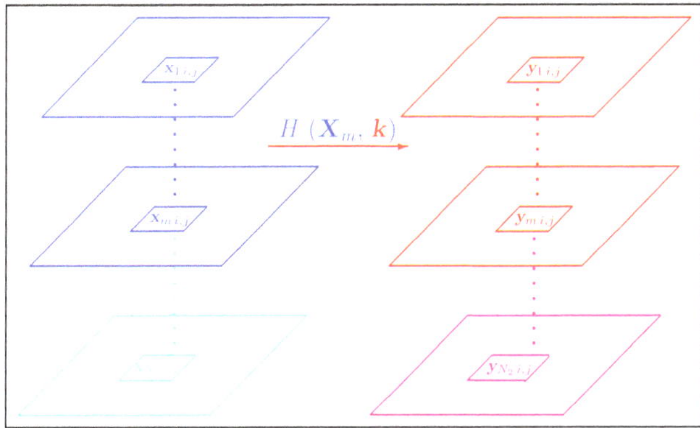

Fig. 1.2.5: Transformation H is a mapping, which maps a subset in the space of space matrices $\mathbb{R}^{m\times M\times N}\times\mathbb{Z}^{+}$ into the space of space matrices $\mathbb{R}^{m\times M\times N}$.

where

$$\begin{aligned}\boldsymbol{X}&=(\boldsymbol{X}^1,\boldsymbol{X}^2,\cdots,\boldsymbol{X}^{N_1})^{\mathrm{T}}\\&=(x_{l\,i,j})_{N_1\times M\times N},\end{aligned}\tag{1.2.75}$$

$$\begin{aligned}\boldsymbol{Y}&=(\boldsymbol{Y}^1,\boldsymbol{Y}^2,\cdots,\boldsymbol{Y}^{N_2})^{\mathrm{T}}\\&=(y_{l\,i,j})_{N_2\times M\times N},\end{aligned}\tag{1.2.76}$$

$$m\le\min\{N_1,N_2\},$$
$$i=1,2,\ldots,M,\quad j=1,2,\ldots,N.$$

In compact forms, CDADS (1.2.73) and (1.2.74) can be written as

$$\boldsymbol{X}(k+1)=F(\boldsymbol{X}(k),\boldsymbol{Y}(k),k),\tag{1.2.77}$$
$$\boldsymbol{Y}(k+1)=G(\boldsymbol{Y}(k),\boldsymbol{X}(k),k),\tag{1.2.78}$$

where

$$F(\boldsymbol{X},\boldsymbol{Y},k)=((f_{1\,i,j}(\boldsymbol{X},\boldsymbol{Y},k))_{M\times N},\ldots,(f_{N_1\,i,j}(\boldsymbol{X},\boldsymbol{Y},k))_{M\times N})^{\mathrm{T}},\tag{1.2.79}$$
$$G(\boldsymbol{Y},\boldsymbol{X},k)=((g_{1\,i,j}(\boldsymbol{Y},\boldsymbol{X},k))_{M\times N},\ldots,(g_{N_2\,i,j}(\boldsymbol{Y},\boldsymbol{X},k))_{M\times N})^{\mathrm{T}}.\tag{1.2.80}$$

Definition 1.21. The non-autonomous bidirectional CDADS defined by (1.2.77) and (1.2.78) are said to be in partial GS (PGS) with respect to a transformation $H:\mathbb{R}^{m\times M\times N}\times\mathbb{Z}^{+}\to\mathbb{R}^{m\times M\times N}$ (see Fig. 1.2.5), if there exists an open subset $B\subset\mathbb{R}^{N_1\times M\times N}\times\mathbb{R}^{N_2\times M\times N}$ such that, for any trajectory $(\boldsymbol{X}(k),\boldsymbol{Y}(k))$ of systems (1.2.77) and (1.2.78) with initial condition $(\boldsymbol{X}(0),\boldsymbol{Y}(0))\in B$, one has

$$\lim_{k\to+\infty}\|H(\boldsymbol{X}_m(k),k)-\boldsymbol{Y}_m(k)\|=0,\tag{1.2.81}$$

where

$$\mathbf{X}_m = (x_{l\,i,j})_{m\times M\times N}, \mathbf{Y}_m = (y_{l\,i,j})_{m\times M\times N}.$$

1.2.4 *PGS of Non-autonomous Bidirectional CCADS*

Definition 1.22. A non-autonomous bidirectional CCADS consists of two parts:

$$\begin{cases} \dot{x}_{1\,i,j}(t) &= f_{1\,i,j}(\mathbf{X}(t), \mathbf{Y}(t), t) \\ \dot{x}_{2\,i,j}(t) &= f_{2\,i,j}(\mathbf{X}(t), \mathbf{Y}(t), t) \\ \quad\vdots \\ \dot{x}_{m\,i,j}(t) &= f_{m\,i,j}(\mathbf{X}(t), \mathbf{Y}(t), t) \\ \quad\vdots \\ \dot{x}_{N_1\,i,j}(t) &= f_{N_1\,i,j}(\mathbf{X}(t), \mathbf{X}(t), t) \end{cases} \tag{1.2.82}$$

and

$$\begin{cases} \dot{y}_{1\,i,j}(t) &= g_{1\,i,j}(\mathbf{Y}(t), \mathbf{X}(t), t) \\ \dot{y}_{2\,i,j}(t) &= g_{2\,i,j}(\mathbf{Y}(t), \mathbf{X}(t), t) \\ \quad\vdots \\ \dot{y}_{m\,i,j}(t) &= g_{m\,i,j}(\mathbf{Y}(t), \mathbf{X}(t), t) \\ \quad\vdots \\ \dot{y}_{N_1\,i,j}(t) &= g_{N_1\,i,j}(\mathbf{X}(t), \mathbf{X}(t), t), \end{cases} \tag{1.2.83}$$

where

$$\begin{aligned} \mathbf{X}(t) &= (\mathbf{X}^1(t), \mathbf{X}^2(t), \cdots, \mathbf{X}^{N_1}(t))^{\mathrm{T}} \\ &= (x_{l\,i,j}(t))_{N_1\times M\times N}, \end{aligned} \tag{1.2.84}$$

$$\begin{aligned} \mathbf{Y}(t) &= (\mathbf{Y}^1(t), \mathbf{Y}^2(t), \cdots, \mathbf{Y}^{N_2}(t))^{\mathrm{T}} \\ &= (y_{l\,i,j}(t))_{N_2\times M\times N}, \end{aligned} \tag{1.2.85}$$

$$m \le \min\{N_1, N_2\},$$
$$i = 1, 2, \ldots, M, \quad j = 1, 2, \ldots, N.$$

In compact forms, CCADS (1.2.82) and (1.2.83) can be written as

$$\dot{\mathbf{X}}(t) = F(\mathbf{X}(t), \mathbf{Y}(t), t), \tag{1.2.86}$$

$$\dot{\mathbf{Y}}(t) = G(\mathbf{Y}(t), \mathbf{X}(t), t), \tag{1.2.87}$$

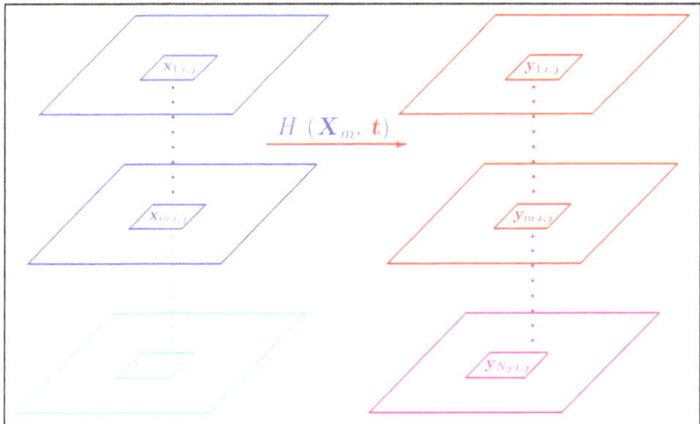

Fig. 1.2.6: Transformation H is a mapping, which maps a subset B in a space matrices $\mathbb{R}^{m\times M\times N}\times\mathbb{R}^{+}$ into a space of space matrices $\mathbb{R}^{m\times M\times N}$.

where

$$F(\boldsymbol{X}(t),\boldsymbol{Y}(t),t)=((f_{1\,i,j}(\boldsymbol{X}(t),\boldsymbol{Y}(t),t))_{M\times N},\dots,\\ (f_{N_1\,i,j}(\boldsymbol{X}(t),\boldsymbol{Y}(t),t))_{M\times N})^{\mathrm{T}}, \tag{1.2.88}$$

$$G(\boldsymbol{Y}(t),\boldsymbol{X}(t))=((g_{1\,i,j}(\boldsymbol{Y}(t),\boldsymbol{X}(t),t))_{M\times N},\dots,\\ (g_{N_2\,i,j}(\boldsymbol{Y}(t),\boldsymbol{X}(t),t))_{M\times N})^{\mathrm{T}}. \tag{1.2.89}$$

Definition 1.23. The bidirectional non-autonomous CCADS defined by (1.2.86) and (1.2.87) are said to be in partial GS (PGS) with respect to a transformation $H:\mathbb{R}^{m\times M\times N}\times\mathbb{R}^{+}\to\mathbb{R}^{m\times M\times N}$ (see Fig. 1.2.6), if there exists an open subset $B\subset\mathbb{R}^{N_1\times M\times N}\times\mathbb{R}^{N_2\times M\times N}$ such that, for any trajectory $(\mathbf{X}(k),\boldsymbol{Y}(k))$ of systems (1.2.86) and (1.2.87) with initial condition $(\boldsymbol{X}(0),\mathbf{Y}(0))\in B$, one has

$$\lim_{t\to+\infty}\|H(\mathbf{X}_m(t),t)-\mathbf{Y}_m(t)\|=0, \tag{1.2.90}$$

where

$$\boldsymbol{X}_m(t)=(x_{l\,i,j}(t))_{m\times M\times N},$$
$$\mathbf{Y}_m(t)=(y_{l\,i,j}(t))_{m\times M\times N}.$$

Remark 1.2.6. As a special case, the transformation $H:\mathbb{R}^{m\times M\times N}\to\mathbb{R}^{m\times M\times N}$ may be independent of the time variable $t\in[0,\infty]$.

1.3 GS Theorems for Discrete and Continuous Arrays of Dynamic Systems

A basic GS problem on coupled discrete arrays of difference equations (see systems (1.2.1) and (1.2.2), (1.2.10) and (1.2.11), (1.2.19) and (1.2.20), (1.2.28) and (1.2.29)), as well as on coupled continuous arrays of differentiable equations (see systems (1.2.37) and (1.2.38), (1.2.46) and (1.2.47), (1.2.55) and (1.2.56), (1.2.64) and (1.2.65)) is as follows:

If the coupled arrays can achieve GS with respect to a transformation H, then what should the general form of the coupled arrays be?

In the following subsections, four general GS theorems are introduced, and seven new GS theorems are derived, on the above-mentioned coupled arrays. These theorems partially answer the above question.

1.3.1 *GS Theorem for Discrete Arrays*

Recall the following GS theorem for CDADS (similar to [Zang and Min (2008)]).

Theorem 1.1. *Let* $\mathbf{X}(k), \mathbf{Y}(k), F(\mathbf{X}(k))$ *and* $G(\mathbf{Y}(k), \mathbf{X}(k))$ *be defined by (1.2.3), (1.2.4), (1.2.7) and (1.2.8), respectively. Suppose that* $H : B \subset \mathbb{R}^{m\times M\times N} \to \mathbb{R}^{m\times M\times N}$ *defined by Definition 1.5 is an invertible transformation. Then, the two systems (1.2.5) and (1.2.6) are in GS with respect to the transformation* $\mathbf{Y}(k) = H(\mathbf{X}_m(k))$ *if, and only if, the function* $G(\mathbf{Y}(k), \mathbf{X}(k))$ *has the following form:*

$$G(\mathbf{Y}(k), \mathbf{X}(k)) = H(F_m(\mathbf{X}(k)) - Q(\mathbf{X}(k), \mathbf{Y}(k)), \tag{1.3.1}$$

where

$$F_m(\mathbf{X}(k)) = ((f_{1\,i,j}(\mathbf{X}(k)))_{M\times N}, \ldots, (f_{m\,i,j}(\mathbf{X}(k)))_{M\times N})^{\mathrm{T}}, \tag{1.3.2}$$

and the function

$$Q(\mathbf{X}(k), \mathbf{Y}(k)) = ((q_{1\,i,j}(\mathbf{X}(k), \mathbf{Y}(k)))_{M\times N}, \ldots, (q_{m\,i,j}(\mathbf{X}(k), \mathbf{Y}(k)))_{M\times N})^{\mathrm{T}}$$

makes the zero solution of the following error equation be asymptotically stable on the open set B *defined by Definition 1.5:*

$$\begin{aligned} \mathbf{e}(k+1) &= H(\mathbf{X}_m(k+1)) - \mathbf{Y}(k+1) \\ &= Q(\mathbf{X}(k), \mathbf{Y}(k)). \end{aligned} \tag{1.3.3}$$

Proof. Denote

$$G(\mathbf{Y}(k), \mathbf{X}(k)) - H(F_m(\mathbf{X}(k))) \\ = -Q(\mathbf{X}(k), \mathbf{Y}(k)).$$

Then,

$$\mathbf{e}(k+1) = H(\mathbf{X}_m(k+1)) - \mathbf{Y}(k+1) \\ = Q(\mathbf{X}(k), \mathbf{Y}(k)). \quad (1.3.4)$$

Therefore, the two dynamic systems (1.2.5) and (1.2.6) are in GS with respect to the transform H if, and only if, the function $Q(\mathbf{X}(k), \mathbf{Y}(k))$ makes the zero solution of the error equation (1.3.4) be zero solution asymptotically stable on the set B defined by Definition 1.5. This completes the proof. □

Remark 1.3.1. In fact, $Q(\mathbf{X}(k), \mathbf{Y}(k))$ can be easily constructed. For example, take $Q(\mathbf{X}(k), \mathbf{Y}(k)) = \lambda \mathbf{e}(k)$ with $\lambda < 1$.

1.3.2 *GS Theorem for Bidirectional Discrete Arrays*

Recall the following GS theorem for bidirectional discrete arrays [Zang *et al.* (2012)].

Theorem 1.2. *Let* $\mathbf{X}(k)$, $\mathbf{Y}(k)$, $F(\mathbf{X}(k), \mathbf{Y}(k))$ *and* $G(\mathbf{X}(k), \mathbf{Y}(k))$ *be defined by (1.2.12), (1.2.13), (1.2.16) and (1.2.17), respectively. Suppose that the transformation* $H : \mathbb{R}^{m\times M\times N} \to \mathbb{R}^{m\times M\times N}$ *is defined by Definition 1.7. Then, the bidirectional CDADSs defined by (1.2.14) and (1.2.15) are in GS with respect to the transformation* $\mathbf{Y}(k) = H(\mathbf{X}_m(k))$ *if, and only if, the function* $G(\mathbf{Y}(k)\mathbf{X}(k))$ *has the following form:*

$$G(\mathbf{Y}(k),\mathbf{X}(k)) = H(F_m(\mathbf{X}(k), \mathbf{Y}(k))) - Q(\mathbf{X}(k),\mathbf{Y}(k)), \quad (1.3.5)$$

where the function

$$Q(\mathbf{X}(k), \mathbf{Y}(k)) = ((q_{1\,i,j}(\mathbf{X}(k), \mathbf{Y}(k)))_{M\times N}, q_{2\,i,j}(\mathbf{X}(k), \mathbf{Y}(k)))_{M\times N}, \\ \ldots, (q_{m\,i,j}(\mathbf{X}(k), \mathbf{Y}(k)))_{M\times N})^T$$

makes the zero solution of the following error equation be asymptotically stable on the open set B *defined by Definition 1.7:*

$$\mathbf{e}(k+1) = H(\mathbf{X}(k+1)) - \mathbf{Y}(k+1) \\ = Q(\mathbf{X}(k), \mathbf{Y}(k)). \quad (1.3.6)$$

Proof. Denote

$$\begin{aligned}&G(\mathbf{Y}(k),\mathbf{X}(k))-H(F_m(\mathbf{X}(k),\mathbf{Y}(k)))\\&=-Q(\mathbf{X}(k),\mathbf{Y}(k)).\end{aligned}$$

Then,

$$\begin{aligned}\mathbf{e}(k+1)&=H(\mathbf{X}_m(k+1))-\mathbf{Y}(k+1)\\&=Q(\mathbf{X}(k),\mathbf{Y}(k)).\end{aligned}\tag{1.3.7}$$

Therefore, the two dynamic systems (1.2.14) and (1.2.15) are in GS with respect to the transform H if and only if the function $Q(\mathbf{X}_m(k),\mathbf{Y}(k))$ makes the zero solution of the error equation (1.3.7) be stable on the set B defined by Definition 1.7.

This completes the proof. □

1.3.3 *GS Theorem for Non-autonomous Discrete Arrays*

Similarly to the case of non-autonomous discrete vector difference systems [Liu *et al.* (2010)], the following result can be established.

Theorem 1.3. *Let* $\mathbf{X}(k)$, $\mathbf{Y}(k)$, $F(\mathbf{X}(k),k)$ *and* $G(\mathbf{X}(k),\mathbf{Y}(k),k)$ *be defined by (1.2.21), (1.2.22), (1.2.25), and (1.2.26), respectively. Suppose that* $H:\mathbb{R}^{m\times M\times N}\times\mathbb{Z}^+\to\mathbb{R}^{m\times M\times N}$ *defined by Definition 1.9. Then, the non-autonomous CDADS defined by (1.2.23) and (1.2.24) are in GS with respect to the transformation* $\mathbf{Y}(k)=H(\mathbf{X}_m(k))$ *if, and only if, the function* $G(\mathbf{Y}(k),\mathbf{X}(k),k)$ *has the following form:*

$$G(\mathbf{Y}(k),\mathbf{X}(k),k)=H(F_m(\mathbf{X}(k),k),k+1)-Q(\mathbf{X}(k),\mathbf{Y}(k),k),\tag{1.3.8}$$

where the function

$$\begin{aligned}Q(\mathbf{X}(k),\mathbf{Y}(k),k)=&((q_{1\,i,j}(\mathbf{X}(k),\mathbf{Y}(k),k))_{M\times N},(q_{2\,i,j}(\mathbf{X}(k),\mathbf{Y}(k),k))_{M\times N},\\&\ldots,(q_{m\,i,j}(\mathbf{X}(k),\mathbf{Y}(k),k))_{M\times N})^{\mathrm{T}}\end{aligned}$$

makes the zero solution of the following error equation be asymptotically stable on the open set B *defined by Definition 1.9:*

$$\begin{aligned}\mathbf{e}(k+1)&=H(\mathbf{X}_m(k+1),k+1)-\mathbf{Y}(k+1)\\&=Q(\mathbf{X}(k),\mathbf{Y}(k),k).\end{aligned}\tag{1.3.9}$$

Proof. Denote

$$\begin{aligned}&G(\mathbf{Y}(k),\mathbf{X}(k),k)-H(F_m(\mathbf{X}(k)),k+1)\\&=-Q(\mathbf{X}(k),\mathbf{Y}(k),k).\end{aligned}$$

Then,

$$\begin{aligned}\mathbf{e}(k+1) &= H(\boldsymbol{X}_m(k+1), k+1) - \boldsymbol{Y}(k+1)\\ &= Q(\boldsymbol{X}(k), \boldsymbol{Y}(k), k).\end{aligned} \tag{1.3.10}$$

Therefore, the two dynamic systems (1.2.23) and (1.2.24) are in GS with respect to the transform H if, and only if, the function $Q(\boldsymbol{X}(k), \mathbf{Y}(k), k)$ makes the zero solution of the error equation (1.3.10) be zero solution asymptotically stable on the set B defined by Definition 1.9. This completes the proof. □

1.3.4 *GS Theorem for Non-autonomous Bidirectional Discrete Arrays*

Similarly to the case of the GS theorem for the non-autonomous discrete vector difference systems [Liu *et al.* (2010)], the following result can be established.

Theorem 1.4. *Let $\mathbf{X}(k)$, $\mathbf{Y}(k)$, $F(\mathbf{X}(k), \mathbf{Y}(k), k)$ and $G(\mathbf{Y}(k), \mathbf{X}(k), k)$ be defined by (1.2.30), (1.2.31), (1.2.34), (1.2.35), respectively. Suppose that $H : \mathbb{R}^{m\times M\times N} \times \mathbb{Z}^+ \to \mathbb{R}^{m\times M\times N}$ is defined by Definition 1.11. Then, the non-autonomous bidirectional CDADSs defined by (1.2.32) and (1.2.33) are in GS with respect to the transformation $\mathbf{Y}(k) = H(\mathbf{X}_m(k))$ if, and only if, the function $G(\mathbf{Y}(k), \mathbf{X}(k), k)$ has the following form:*

$$\begin{aligned}G(\mathbf{Y}(k), \mathbf{X}(k), k) &= H(F_m(\mathbf{X}(k), \mathbf{Y}(k), k), k+1)\\ &\quad -Q(\mathbf{X}(k), \mathbf{Y}(k), k),\end{aligned} \tag{1.3.11}$$

where the function

$$\begin{aligned}Q(\mathbf{X}(k), \mathbf{Y}(k), k) = ((&q_{1\,i,j}(\mathbf{X}(k), \mathbf{Y}(k), k))_{M\times N},\\ &(q_{2\,i,j}(\mathbf{X}(k), \mathbf{Y}(k), k))_{M\times N}, \ldots,\\ &(q_{m\,i,j}(\mathbf{X}(k), \mathbf{Y}(k), k))_{M\times N})^{\mathrm{T}}\end{aligned}$$

makes the zero solution of the following error equation be zero solution asymptotically stable on the open set B defined by the Definition 1.11:

$$\begin{aligned}\mathbf{e}(k+1) &= H(\mathbf{X}_m(k+1)) - \mathbf{Y}(k+1)\\ &= Q(\mathbf{X}(k), \mathbf{Y}(k), k).\end{aligned} \tag{1.3.12}$$

Proof. Denote

$$\begin{aligned}&G(\boldsymbol{Y}(k), \boldsymbol{X}(k), k) - H(F_m(\boldsymbol{X}(k), \boldsymbol{Y}(k), k), k+1)\\ &\quad = -Q(\boldsymbol{X}(k), \boldsymbol{Y}(k), k).\end{aligned}$$

Then,

$$\begin{aligned}\mathbf{e}(k+1) &= H(\mathbf{X}_m(k+1), k+1) - \mathbf{Y}(k+1)\\ &= H(F_m(\mathbf{X}(k), \mathbf{Y}(k), k), k+1)\\ &\quad -H(F_m(\mathbf{X}(k), \mathbf{Y}(k), k+1) + Q(\mathbf{X}, \mathbf{Y}, k)\\ &= Q(\mathbf{X}, \mathbf{Y}, k).\end{aligned}$$

Therefore, the two dynamic systems (1.2.32) and (1.2.33) are in GS with respect to the transformation H if, and only if, the function $Q(\mathbf{X}(k), \mathbf{Y}(k), k)$ makes the zero solution of the error equation (1.3.12) be asymptotically stable on the set B defined by Definition 1.11. This completes the proof. □

1.3.5 *GS Theorem for an Array of Differential Systems*

Recall the following GS theorem for an array of differential systems [Min and Zang (2009)].

Theorem 1.5. *Let $\mathbf{X}(t)$, $\mathbf{Y}(t)$, $F(\mathbf{X}(t))$ and $G(\mathbf{X}(t), \mathbf{Y}(t))$ be defined by (1.2.39), (1.2.40), (1.2.43), and (1.2.44), respectively. Then, the two systems (1.2.41) and (1.2.42) are in GS with respect to the transformation $\mathbf{Y} = H(\mathbf{X}_m)$ defined by Definition 1.13 (also see Fig. 1.3.1) if, and only if, the function $G(\mathbf{Y}(t), \mathbf{X}(t))$ has the following form:*

$$\begin{aligned}G_{l\,i,j}(\mathbf{Y}(t), \mathbf{X}(t)) &= \sum_{l'=1}^{m}\sum_{i'=1}^{M}\sum_{j'=1}^{N} \frac{\partial h_{l\,i,j}(\mathbf{X}_m(t))}{\partial x_{l'\,i',j'}}\\ &\quad \times f_{l'\,i',j'}(\mathbf{X}(t) - q_{l\,i,j}(\mathbf{Y}(t), \mathbf{X}(t)), \qquad (1.3.13)\\ &\quad l = 1, \cdots, m, i = 1, \cdots, M,\\ &\quad j = 1, \cdots, N,\end{aligned}$$

where the function $q_{l\,i,j}(\mathbf{X}(t), \mathbf{Y}(t))$ make the zero solution of the error equation

$$\begin{aligned}\frac{d\mathbf{e}}{dt} &= \frac{d(H(\mathbf{X}_m(t)) - \mathbf{Y}\ (t))}{dt}\\ &= (q_{l\,i,j}(\mathbf{Y}, \mathbf{X}, t))_{n\times M\times N} \qquad (1.3.14)\end{aligned}$$

be asymptotically stable on the open set B defined by Definition 1.13.

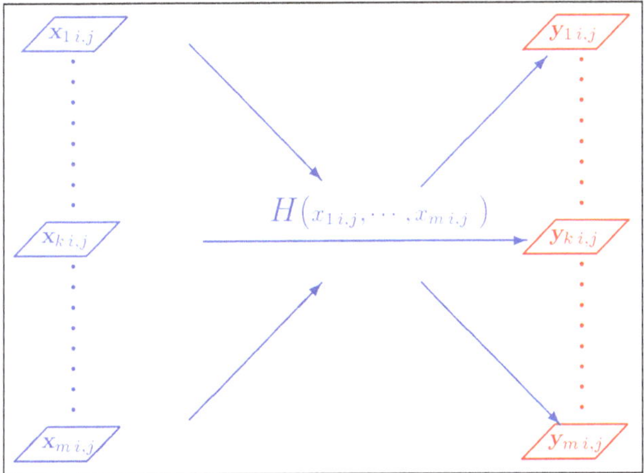

Fig. 1.3.1: For each fixed (i, j), transformation H is a mapping,which maps an m-dimensional vector $[x_{1\,i,j}, \cdots, x_{m,\,i,j}] \in \mathbb{R}^m$ into $\mathbb{R}^m$.

1.3.6 *GS Theorem for a Bidirectional Array of Differential Systems*

Recall the following GS theorem for a bidirectional array of differential systems [Zang *et al.* (2012)].

Theorem 1.6. *Let* $\mathbf{X}(t)$*,* $\mathbf{Y}(t)$*,* $F(\mathbf{X}(t), \mathbf{Y}(t))$ *and* $G(\mathbf{Y}(t), \mathbf{X}(t))$ *be defined by (1.2.48), (1.2.49), (1.2.52) and (1.2.53), respectively. Suppose that* $H : \mathbb{R}^{m\times M\times N} \to \mathbb{R}^{m\times M\times N}$ *defined by (1.15), and Fig. 1.3.1 is continuously differentiable. Then, the bidirectional CCADS defined by (1.2.50) and (1.2.51) are in GS with respect to the transformation* H *if, and only if, the function* $G(\mathbf{Y}, \mathbf{X})$ *has the following forms:*

$$G(\mathbf{Y}, \mathbf{X}) = \Bigg(\frac{\partial H(x_{1\,i,j}, \cdots, x_{m\,i,j})}{\partial \mathbf{X}} [f_{1\,i,j}(\mathbf{X}, \mathbf{Y}), \cdots, f_{m\,i,j}(\mathbf{X}, \mathbf{Y})]^T\Bigg)_{M\times N} - Q(\mathbf{X}, \mathbf{Y}), \tag{1.3.15}$$

where

$$\frac{\partial H(x_{1\,i,j},\cdots,x_{m\,i,j})}{\partial \mathbf{X}} = \begin{bmatrix} \frac{\partial h_1}{\partial x_{1\,i,j}} & \frac{\partial h_1}{\partial x_{2\,i,j}} & \cdots & \frac{\partial h_1}{\partial x_{m\,i,j}} \\ \frac{\partial h_2}{\partial x_{1\,i,j}} & \frac{\partial h_2}{\partial x_{2\,i,j}} & \cdots & \frac{\partial h_2}{\partial x_{m\,i,j}} \\ \vdots & \vdots & \vdots & \vdots \\ \frac{\partial h_m}{\partial x_{1\,i,j}} & \frac{\partial h_m}{\partial x_{2\,i,j}} & \cdots & \frac{\partial h_m}{\partial x_{m\,i,j}} \end{bmatrix},$$

and $Q(\mathbf{X},\mathbf{Y})$ *makes the zero solution of the error equation*

$$\begin{aligned} \dot{\mathbf{E}} &= H(\mathbf{X}_m) - \frac{d(\mathbf{Y})}{dt} \\ &= Q(\mathbf{X},\mathbf{Y}) \\ &= [q_{1\,i,j}(\mathbf{X},\mathbf{Y})_{M\times N},\cdots,q_{m\,i,j}(\mathbf{X},\mathbf{Y})_{M\times N}]^{\mathrm{T}} \end{aligned} \tag{1.3.16}$$

be asymptotically stable on the open set B defined by Definition 1.15, where $\mathbf{E} = H(\mathbf{X}_m) - \mathbf{Y}$.

Similarly to the theorems established in [Ji *et al.* (2008); Min and Zang (2009)], Theorem 1.6 can be extended to the following.

Theorem 1.7. *Suppose that $H : \mathbb{R}^{m\times M\times N} \to \mathbb{R}^{m\times M\times N}$ defined by Definition 1.15 (also see Fig. 1.2.2) is continuously differentiable. Let $\mathbf{X}(t)$, $\mathbf{Y}(t)$, $F(\mathbf{X}(t),\mathbf{Y}(t))$ and $G(\mathbf{X}(t),\mathbf{Y}(t))$ be defined by (1.2.48), (1.2.49), (1.2.52) and (1.2.53), respectively. Then, the bidirectional CCADS defined by (1.2.50) and (1.2.51) are in GS with respect to the transformation $\mathbf{Y} = H(\mathbf{X}_m)$ if, and only if, the function $G(\mathbf{Y},\mathbf{X})$ has the following form:*

$$G_{l\,i,j}(\mathbf{Y},\mathbf{X}) = \sum_{l'=1}^{m}\sum_{i'=1}^{M}\sum_{j'=1}^{N} \frac{\partial h_{l\,i,j}(\mathbf{X}_m)}{\partial x_{l'\,i',j'}} f_{l'\,i',j'}(\mathbf{X},\mathbf{Y}) - q_{l\,i,j}(\mathbf{X},\mathbf{Y}) \tag{1.3.17}$$
$$l = 1,\cdots,m,\ i = 1,\cdots,M,\ j = 1,\cdots,N,$$

where the function $q_{l\,i,j}(\mathbf{X},\mathbf{Y})$ make the zero solution of the error equation

$$\begin{aligned} \frac{d\mathbf{e}}{dt} &= \frac{d(H(\mathbf{X}_m) - \mathbf{Y})}{dt} \\ &= (q_{l\,i,j}(\mathbf{X},\mathbf{Y}))_{m\times M\times N} \end{aligned} \tag{1.3.18}$$

be asymptotically stable on the open set B defined by Definition 1.15.

Proof. $\Rightarrow$ Since H is continuously differentiable, one has

$$G_{l\,i,j}(\boldsymbol{Y},\boldsymbol{X})=\sum_{l'=1}^{m}\sum_{i'=1}^{M}\sum_{j'=1}^{N}\frac{\partial h_{l\,i,j}(\boldsymbol{X}_m)}{\partial x_{l'\,i',j'}}f_{l'\,i',j'}(\boldsymbol{X},\boldsymbol{Y})-q_{l\,i,j}(\boldsymbol{Y},\boldsymbol{X}).$$

Let

$$\mathbf{e}=H(\boldsymbol{X}_m)-\boldsymbol{Y}.$$

Then, the error equation has the following form:

$$\begin{aligned}\frac{d\mathbf{e}}{dt}&=\left(\frac{dh_{l\,i,j}(\boldsymbol{X}_m)}{dt}\right)_{m\times M\times N}-\frac{d\boldsymbol{Y}}{dt}\\&=\left(\sum_{l'=1}^{m}\sum_{i'=1}^{M}\sum_{j'=1}^{N}\frac{\partial h_{l\,i,j}(\boldsymbol{X}_m)}{\partial x_{l'\,i',j'}}\dot{x}_{l'\,i',j'}\right)_{m\times M\times N}-(G_{l\,i,j}(\boldsymbol{Y},\boldsymbol{X}))_{m\times M\times N}\\&=\left(\sum_{l'=1}^{m}\sum_{i'=1}^{M}\sum_{j'=1}^{N}\frac{\partial h_{l\,i,j}(\boldsymbol{X}_m)}{\partial x_{l'\,i',j'}}f_{l'\,i',j'}(\boldsymbol{X},\boldsymbol{Y})\right)_{m\times M\times N}-(G_{l\,i,j}(\boldsymbol{Y},\boldsymbol{X}))_{m\times M\times N}\\&=(q_{l\,i,j}(\boldsymbol{X},\boldsymbol{Y}))_{m\times M\times N}.\end{aligned}\tag{1.3.19}$$

By assumption, CCADS (1.2.50) and (1.2.51) are in GS with respect to the transformation H. Therefore, $q_{l\,i,j}(\boldsymbol{X},\boldsymbol{Y})$ ensures the trajectory of the system (1.3.19) to tend to zero asymptotically from any initial condition in the set B defined by Definition 1.15.

$\Leftarrow$ If the function $G_{l\,i,j}(\boldsymbol{Y},\boldsymbol{X})$ can be represented by equation (1.3.17), and $(q_{l\,i,j}(\boldsymbol{X},\boldsymbol{Y}))_{m\times M\times N}$ makes the zero solution of equation (1.3.18) be asymptotically stable on the set B defined by Definition 1.15, then the CCADS (1.2.50) and (1.2.51) are in GS with respect to the transformation $\boldsymbol{Y}=H(\boldsymbol{X}_m)$.

This completes the proof. □

1.3.7 *GS Theorem for a Non-autonomous Array of Differential Systems*

Similarly to the theorems established in [Liu *et al.* (2007); Min and Zang (2009)], the following theorem can be proved.

Theorem 1.8. *Suppose that* $H:\mathbb{R}^{m\times M\times N}\times\mathbb{R}^{+}\to\mathbb{R}^{m\times M\times N}$ *defended by Definition 1.17 is continuously differentiable. Let* $\mathbf{X}(t),\mathbf{Y}(t),F(\mathbf{X}(t),t)$ *and* $G(\mathbf{X}(t),\mathbf{Y}(t),t)$ *be defined by (1.2.57), (1.2.58), (1.2.61), (1.2.62),* *respectively. Then, the non-autonomous CCADS defined by (1.2.59) and (1.2.60) are in GS with respect to the transformation* $\mathbf{Y} = H(\mathbf{X}_m, t)$ *if, and only if, the function* $G(\mathbf{Y}, \mathbf{X}, t)$ *has the following form:*

$$\begin{aligned} G_{l\,i,j}(\mathbf{Y}, \mathbf{X}, t) = & \sum_{l'=1}^{m}\sum_{i'=1}^{M}\sum_{j'=1}^{N} \frac{\partial h_{l\,i,j}(\mathbf{X}_m, t)}{\partial x_{l'\,i',j'}} f_{l'\,i',j'}(\mathbf{X}, t) \\ & + \frac{\partial h_{l\,i,j}(\mathbf{X}_m, t)}{\partial t} - q_{l\,i,j}(\mathbf{Y}, \mathbf{X}, t), \qquad (1.3.20) \\ & l = 1, \cdots, m, i = 1, \cdots, M, \\ & j = 1, \cdots, N, \end{aligned}$$

where the function $q_{l\,i,j}(\mathbf{X}, \mathbf{Y}, t)$ *make the error solution of the error equation*

$$\begin{aligned} \frac{d\mathbf{e}}{dt} &= \frac{d(H(\mathbf{X}_m, t) - \mathbf{Y})}{dt} \\ &= (q_{l\,i,j}(\mathbf{Y}, \mathbf{X}, t))_{m\times M\times N} \qquad (1.3.21) \end{aligned}$$

be asymptotically stable on the open set B defined by Definition 1.17.

Proof. ⇒ Since $H(\mathbf{X}_m, t)$ is continuously differentiable, one has

$$\begin{aligned} G_{l\,i,j}(\mathbf{Y}, \mathbf{X}, t) = & \sum_{l'=1}^{m}\sum_{i'=1}^{M}\sum_{j'=1}^{N} \frac{\partial h_{l\,i,j}(\mathbf{X}_m, t)}{\partial x_{l'\,i',j'}} f_{l'\,i',j'}(\mathbf{X}, t) \\ & + \frac{\partial h_{l\,i,j}(\mathbf{X}_m, t)}{\partial t} - q_{l\,i,j}(\mathbf{Y}, \mathbf{X}, t). \qquad (1.3.22) \end{aligned}$$

Let

$$\mathbf{e} = H(\mathbf{X}_m, t) - \mathbf{Y}.$$

Then, the error equation has the following form:

$$\begin{aligned} \frac{d\mathbf{e}}{dt} &= \left(\frac{dh_{l\,i,j}(\mathbf{X}_m, t)}{dt}\right)_{m\times M\times N} - \frac{d\mathbf{Y}}{dt} \\ &= \left(\sum_{l'=1}^{m}\sum_{i'=1}^{M}\sum_{j'=1}^{N} \frac{\partial h_{l\,i,j}(\mathbf{X}_m, t)}{\partial x_{l'\,i',j'}} \dot{x}_{l'\,i',j'}\right)_{m\times M\times N} \\ &\quad + \left(\frac{\partial h_{l\,i,j}(\mathbf{X}_m, t)}{\partial t}\right)_{m\times M\times N} - (G_{l\,i,j}(\mathbf{Y}, \mathbf{X}, t))_{m\times M\times N} \\ &= \left(\sum_{l'=1}^{m}\sum_{i'=1}^{M}\sum_{j'=1}^{N} \frac{\partial h_{l\,i,j}(\mathbf{X}_m, t)}{\partial x_{l'\,i',j'}} f_{l'\,i',j'}(\mathbf{X}, t)\right)_{m\times M\times N} \\ &\quad + \left(\frac{\partial h_{l\,i,j}(\mathbf{X}_m, t)}{\partial t}\right)_{m\times M\times N} - (G_{l\,i,j}(\mathbf{Y}, \mathbf{X}, t))_{m\times M\times N} \\ &= (q_{l\,i,j}(\mathbf{X}, \mathbf{Y}, t))_{m\times M\times N}. \qquad (1.3.23) \end{aligned}$$

By assumption, the two CCADS (1.2.59) and (1.2.60) are in GS with respect to the transformation H. Therefore, $q_{l\,i,j}(\mathbf{X}, \mathbf{Y}, t)$ ensures the trajectory of the error equation (1.3.21) to tend to zero asymptotically for any initial condition in the set B defined by Definition 1.17.

$\Leftarrow$ If the function $G_{l\,i,j}(\mathbf{Y}, \mathbf{X}, t)$ can be represented by equation (1.3.20), and $(q_{l\,i,j}(\mathbf{X}, \mathbf{Y}, t))_{m\times M\times N}$ makes the zero solution of the error equation (1.3.21) be asymptotically stable on the set B defined by Definition 1.17, then the two CCADS (1.2.59) and (1.2.60) are in GS with respect to the transformation $H(\mathbf{X}_m, t) = \mathbf{Y}$.

This completes the proof. □

1.3.8 *GS Theorem for a Non-autonomous Bidirectional Array of Differential Systems*

Similarly to the theorems established in [Min and Zang (2009); Liu *et al.* (2007)], the following result can be proved.

Theorem 1.9. *Suppose that $H : \mathbb{R}^{m\times M\times N} \times \mathbb{R}^+ \to \mathbb{R}^{m\times M\times N}$ defined by Definition 1.19 is continuously differentiable. Let $\mathbf{X}(k)$, $\mathbf{Y}(k)$, $F(\mathbf{X}(k), \mathbf{Y}(k), t)$ and $G(\mathbf{X}(k), \mathbf{Y}(k), t)$ be defined by (1.2.66), (1.2.67), (1.2.70) and (1.2.71), respectively. Then, the nonautonomous bidirectional CCADS (1.2.68) and (1.2.69) are in GS with respect to the transformation $\mathbf{Y} = H(\mathbf{X}_m, t)$ if, and only if, the function $G(\mathbf{Y}, \mathbf{X}, t)$ has the following form:*

$$\begin{aligned} G_{l\,i,j}(\mathbf{Y}, \mathbf{X}, t) = {} & \sum_{l'=1}^{n}\sum_{i'=1}^{M}\sum_{j'=1}^{N} \frac{\partial h_{l\,i,j}(\mathbf{X}_m, t)}{\partial x_{l'\,i',j'}} \\ & \times f_{l'\,i',j'}(\mathbf{X}, \mathbf{Y}, t) + \frac{\partial h_{l\,i,j}(\mathbf{X}_m, t)}{\partial t} \\ & - q_{l\,i,j}(\mathbf{Y}, \mathbf{X}, t), \\ & l = 1, \cdots, n, i = 1, \cdots, M, \\ & j = 1, \cdots, N, \end{aligned} \tag{1.3.24}$$

where the function $q_{l\,i,j}(\mathbf{X}, \mathbf{Y}, t)$ make the zero solution of the error equation

$$\begin{aligned} \frac{d\mathbf{e}}{dt} &= \frac{d(H(\mathbf{X}_m, t) - \mathbf{Y})}{dt} \\ &= (q_{l\,i,j}(\mathbf{Y}, \mathbf{X}, t))_{m\times M\times N} \end{aligned} \tag{1.3.25}$$

be asymptotically stable on the open set B defined by Definition 1.19.

Proof. $\Rightarrow$ Since $H(\mathbf{X}_m, t)$ is continuously differentiable, one has

$$G_{l\,i,j}(\mathbf{Y}, \mathbf{X}, t) = \sum_{l'=1}^{n}\sum_{i'=1}^{M}\sum_{j'=1}^{N} \frac{\partial h_{l\,i,j}(\mathbf{X}_m)}{\partial x_{l'\,i',j'}} \times f_{l'\,i',j'}(\mathbf{X}, \mathbf{Y}, t) + \frac{\partial h_{l\,i,j}(\mathbf{X}_m, t)}{\partial t} - q_{l\,i,j}(\mathbf{Y}, \mathbf{X}, t).$$

Let

$$\mathbf{e} = H(\mathbf{X}_m, t) - \mathbf{Y}.$$

Then, the error equation has the following form:

$$\begin{aligned}
\frac{d\mathbf{e}}{dt} &= \left(\frac{dh_{l\,i,j}(\mathbf{X}_m, t)}{dt}\right)_{m\times M\times N} - \frac{d\mathbf{Y}}{dt} \\
&= \left(\sum_{l'=1}^{m}\sum_{i'=1}^{M}\sum_{j'=1}^{N} \frac{\partial h_{l\,i,j}(\mathbf{X}_m)}{\partial x_{l'\,i',j'}} \dot{x}_{l'\,i',j'}\right)_{m\times M\times N} \\
&\quad + \left(\frac{\partial h_{l\,i,j}(\mathbf{X}_m, t)}{\partial t}\right)_{m\times M\times N} - (G_{l\,i,j}(\mathbf{Y}, \mathbf{X}, t))_{m\times M\times N} \\
&= \left(\sum_{l'=1}^{m}\sum_{i'=1}^{M}\sum_{j'=1}^{N} \frac{\partial h_{l\,i,j}(\mathbf{X}_m)}{\partial x_{l'\,i',j'}} \times f_{l'\,i',j'}(\mathbf{X}, \mathbf{Y}, t)\right)_{m\times M\times N} \\
&\quad + \left(\frac{\partial h_{l\,i,j}(\mathbf{X}_m, t)}{\partial t}\right)_{m\times M\times N} - (G_{l\,i,j}(\mathbf{Y}, \mathbf{Y}, t))_{m\times M\times N} \\
&= (q_{l\,i,j}(\mathbf{X}, \mathbf{Y}, t))_{m\times M\times N}\,.
\end{aligned}$$

By assumption, the CCADS (1.2.68) and (1.2.69) are in GS with respect to the transformations H. Therefore, $q_{l\,i,j}(\mathbf{X}, \mathbf{Y}, t)$ make the zero solution of the error equation (1.3.25) be asymptotically stable for any initial condition in the set B defined by Definition 1.19.

$\Leftarrow$ If the function $G_{l\,i,j}(\mathbf{Y}, \mathbf{X}, t)$ can be represented by equation (1.3.24), and $(q_{l\,i,j}(\mathbf{X}, \mathbf{Y}, t))_{m\times M\times N}$ makes the zero solution of equation (1.3.25) be asymptotically stable on the set B defined by Definition 1.19, then condition (1.2.73) holds. This means the two CCADS (1.2.68) and (1.2.69) are in GS with respect to the transformation $H(\mathbf{X}_m, t) = \mathbf{Y}$.

This completes the proof. □

1.3.9 *PGC Theorem for Non-autonomous Bidirectional CDADS*

Non-autonomous bidirectional coupled discrete arrays are a kind of the most complex and general discrete networks. The partial GC (PGC) theory for them is of common interest.

Similarly to the partial GS Theorem 1.10 for non-autonomous bidirectional coupled discrete arrays (also see Theorem 10 given in [Min and Chen (2013)]), we have the following result.

Theorem 1.10. *Let* $\mathbf{X}(k)$, $\mathbf{Y}(k)$, $F(\mathbf{X}(k), \mathbf{Y}(k), k)$ *and* $G(\mathbf{X}(k), \mathbf{Y}(k), k)$ *be defined by (1.2.75), (1.2.76), (1.2.79), and (1.2.80), respectively. Suppose that the transformation* $H : \mathbb{R}^{m\times M\times N} \times \mathbb{Z}^{+} \to \mathbb{R}^{m\times M\times N}$ *is defined by Definition 1.21. Then, the non-autonomous bidirectional CADDS defined by (1.2.77) and (1.2.78) are in partial GS (PGS) with respect to the transformation* $\mathbf{Y}_m(k) = H(\mathbf{X}_m(k), k)$ *if, and only if, the function*

$$G_m(\mathbf{Y}(k), \mathbf{X}(k), k) = (g_{l\,i,j}(\mathbf{Y}(k), \mathbf{X}(k), k))_{m\times M\times M} \tag{1.3.26}$$

given in (1.2.80) has the following form:

$$\begin{aligned} G_m(\mathbf{Y}(k), \mathbf{X}(k), k) = H(&F_m(\mathbf{X}(k), \mathbf{Y}(k), k), k+1) \\ &-Q(\mathbf{X}(k), \mathbf{Y}(k), k), \end{aligned} \tag{1.3.27}$$

where the function

$$\begin{aligned} Q(\mathbf{X}(k), \mathbf{Y}(k), k) = [&(q_{1\,i,j}(\mathbf{X}(k), \mathbf{Y}(k), k))_{M\times N}, (q_{2\,i,j}(\mathbf{X}(k), \mathbf{Y}(k), k))_{M\times N}, \\ &\ldots, (q_{m\,i,j}(\mathbf{X}(k), \mathbf{Y}(k), k))_{M\times N}]^{\mathrm{T}} \end{aligned} \tag{1.3.28}$$

makes the zero solution of the following error equation be zero asymptotically stable on the open set B *defined by the Definition 1.21:*

$$\begin{aligned} \mathbf{e}(k+1) &= H(\mathbf{X}_m(k+1), k+1) - \mathbf{Y}_m(k+1) \\ &= Q(\mathbf{X}(k), \mathbf{Y}(k), k). \end{aligned} \tag{1.3.29}$$

Proof. Denote

$$\begin{aligned} &G_m(\boldsymbol{Y}(k), \boldsymbol{X}(k), k) - H(F_m(\boldsymbol{X}(k), \boldsymbol{Y}(k), k+1)) \\ &\quad = -Q(\boldsymbol{X}(k), \boldsymbol{Y}(k), k). \end{aligned} \tag{1.3.30}$$

Then,

$$\begin{aligned} \mathbf{e}(k+1) &= H(\boldsymbol{X}_m(k+1), k+1) - \boldsymbol{Y}_m(k+1). \\ &= Q(\boldsymbol{X}(k), \boldsymbol{Y}(k), k). \end{aligned} \tag{1.3.31}$$

Therefore, the two systems (1.2.77) and (1.2.78) are in GS with respect to the transformation H defined by Definition 1.21 if, and only if, the function $Q(\mathbf{X}(k), \mathbf{X}(k), k)$ makes the zero solution of the error equation (1.3.29) be asymptotically stable.

This completes the proof. □

Remark 1.3.2. In fact, $Q(\mathbf{X}(k), \mathbf{Y}(k), k)$ can be easily constructed. For example, take $Q(\mathbf{X}(k), \mathbf{Y}(k), k) = \lambda \mathbf{e}(k)$ with $\lambda < 1$.

1.3.10 *PGS Theorem for Non-autonomous Bidirectional CCADS*

Non-autonomous CCADS are a kind of the most complex and general continuous networks. The partial GS (PGS) theory for them is of common interest. Similarly to Theorem 1.9, the following result can be proved.

Theorem 1.11. *Suppose that* $H : \mathbb{R}^{m\times M\times N} \times \mathbb{R}^{+} \to \mathbb{R}^{m\times M\times N}$ *is continuously differentiable. Let* $\mathbf{X}(t), \mathbf{Y}(t), F(\mathbf{X}(t), \mathbf{Y}(t), t)$ *and* $G(\mathbf{X}(t), \mathbf{Y}(t), t)$ *be defined by (1.2.84), (1.2.85), (1.2.88), and (1.2.89), respectively. Then, the non-autonomous bidirectional CCADS defined by (1.2.86) and (1.2.87) are in partial GS (PGS) with respect to the transformation* $\mathbf{Y}_m(t) = H(\mathbf{X}_m(t), t)$ *defined by Definition 1.23 if, and only if, the function* $G(\mathbf{Y}, \mathbf{X}, t)$ *has the following form:*

$$\begin{aligned} G_{l\,i,j}(\mathbf{Y}, \mathbf{X}, t) = &\sum_{l'=1}^{m}\sum_{i'=1}^{M}\sum_{j'=1}^{N} \frac{\partial h_{l\,i,j}(\mathbf{X}_m, t)}{\partial x_{l'\,i',j'}} \times f_{l'\,i',j'}(\mathbf{X}, \mathbf{Y}, t) \\ &+ \frac{\partial h_{l\,i,j}(\mathbf{X}_m, t)}{\partial t} - q_{l\,i,j}(\mathbf{Y}, \mathbf{X}, t), \qquad (1.3.32) \\ l = 1, \cdots, m, i &= 1, \cdots, M, j = 1, \cdots, N, \end{aligned}$$

where the function $q_{l\,i,j}(\mathbf{Y}, \mathbf{X}, t)$ *make the zero solution of the error equation*

$$\begin{aligned} \frac{d\mathbf{e}}{dt} &= \frac{d(H(\mathbf{X}_m, t) - \mathbf{Y}_m)}{dt} \\ &= (q_{l\,i,j}(\mathbf{Y}, \mathbf{X}, t))_{m\times M\times N} \qquad (1.3.33) \end{aligned}$$

be asymptotically stable on the open set B *defined by Definition 1.23.*

Proof. ⇒ Since $H(\mathbf{X}_m, t)$ is continuously differentiable, one has

$$\begin{aligned} G_{l\,i,j}(\mathbf{Y}, \mathbf{X}, t) = &\sum_{l'=1}^{m}\sum_{i'=1}^{M}\sum_{j'=1}^{N} \frac{\partial h_{l\,i,j}(\mathbf{X}_m, t)}{\partial x_{l'\,i',j'}} \times f_{l'\,i',j'}(\mathbf{X}, \mathbf{Y}, t) \\ &+ \frac{\partial h_{l\,i,j}(\mathbf{X}_m, t)}{\partial t} - q_{l\,i,j}(\mathbf{Y}, \mathbf{X}, t). \end{aligned}$$

Let

$$\mathbf{e} = H(\mathbf{X}_m, t) - \mathbf{Y}_m.$$

Then, the error equation has the following form:

$$\begin{aligned}
\frac{d\mathbf{e}}{dt} &= \left(\frac{dh_{l\,i,j}(\mathbf{X}_m, t)}{dt}\right)_{m\times M\times N} - \frac{d\mathbf{Y}_m}{dt} \\
&= \left(\sum_{l'=1}^{m}\sum_{i'=1}^{M}\sum_{j'=1}^{N}\frac{\partial h_{l\,i,j}(\mathbf{X}_m, t)}{\partial x_{l'\,i',j'}}\dot{x}_{l'\,i',j'}\right)_{m\times M\times N} \\
&\quad + \left(\frac{\partial h_{l\,i,j}(\mathbf{X}_m, t)}{\partial t}\right)_{m\times M\times N} - (G_{l\,i,j}(\mathbf{Y}, \mathbf{X}, t))_{m\times M\times N} \\
&= \left(\sum_{l'=1}^{m}\sum_{i'=1}^{M}\sum_{j'=1}^{N}\frac{\partial h_{l\,i,j}(\mathbf{X}_m, t)}{\partial x_{l'\,i',j'}} \times f_{l'\,i',j'}(\mathbf{X}, \mathbf{Y}, t)\right)_{m\times M\times N} \\
&\quad + \left(\frac{\partial h_{l\,i,j}(\mathbf{X}_m, t)}{\partial t}\right)_{m\times M\times N} - (G_{l\,i,j}(\mathbf{Y}, \mathbf{Y}, t))_{m\times M\times N} \\
&= (q_{l\,i,j}(\mathbf{Y}, \mathbf{X}, t))_{m\times M\times N}\,.
\end{aligned}$$

By assumption, the two CCADS (1.2.86) and (1.2.87) are in PGS via the transformations H. Therefore, $q_{l\,i,j}(\mathbf{Y}, \mathbf{X}, t)$ makes the zero solution of the error equation (1.3.33) be asymptotically stable for any initial condition in the set B defined by Definition 1.23.

$\Leftarrow$ If the function $G_{l\,i,j}(\mathbf{Y}, \mathbf{X}, t)$ can be represented by equation (1.3.32), and $(q_{l\,i,j}(\mathbf{Y}, \mathbf{X}, t))_{m\times M\times N}$ makes the zero solution of equation (1.3.33) be asymptotically stable on the set B defined by Definition 1.23, then condition (1.2.90) holds. This means that the two CCADS (1.2.86) and (1.2.87) are in GS with respect to the transformation $H(\mathbf{X}_m, t) = \mathbf{Y}_m$.

This completes the proof. □

Remark 1.3.3. In fact, $Q(\mathbf{Y}(k), \mathbf{X}(k), k) = (q_{l\,i,j}(\mathbf{Y}, \mathbf{X}, t))_{m\times M\times N}$ can be easily constructed. For example, take $Q(\mathbf{Y}(k),\ \mathbf{X}(k), k) = \lambda\mathbf{e}(k)$ where $\lambda < 0$.

1.4 Application of GS Theorems

1.4.1 *Application of GS Theorem to CDADS*

In this subsection, an application of Theorem 1.1 to coupled GS systems is discussed.

Based on the hyperchaotic Chen system proposed by Jia *et al.* [Jia *et al.* (2010)], a discrete Chen cellular neural network (CNN) is introduced as follows:

$$\begin{cases} x_{1\,i,j}(k+1) = x_{1\,i,j}(k) + \epsilon a[x_{2\,i,j}(k) - x_{1\,i,j}(k)] \\ x_{2\,i,j}(k+1) = x_{2\,i,j}(k) + \epsilon[4x_{1,i,j}(k) + cx_{2\,i,j}(k) \\ \qquad\qquad -10x_{1\,i,j}(k)x_{3\,i,j}(k) + 4x_{4\,i,j}(k)] \\ x_{3\,i,j}(k+1) = x_{3\,i,j}(k) + \epsilon[x_{2\,i,j}^2(k) - bx_{3\,i,j}(k)] \\ x_{4\,i,j}(k+1) = x_{4\,i,j}(k) - \epsilon dx_{1\,i,j}(k) \\ \qquad\qquad +D_2[x_{3i+1,j} + x_{3i-1,j} \\ \qquad\qquad +x_{3i,j+1} + x_{3i,j-1} - 4x_{4i,j}], \\ \qquad\qquad i,j = 1,2,\cdots,21, \end{cases} \tag{1.4.1}$$

where

$$a = 35, b = 3, c = 21, d = 2, \epsilon = 0.001, D = 0.01.$$

Written in a compact form, it is

$$\mathbf{X}(k+1) = F(\mathbf{X}(k)). \tag{1.4.2}$$

Now, select the following initial conditions:

$$(x_{l\,i,j}(0))_{21\times21} = 0.1(1 + rand(21,21)), \tag{1.4.3}$$

where $rand(21,21)$ is a matlab command, which returns a 21×21 matrix containing pseudo-random values drawn from a uniform distribution on the unit interval.

The chaotic orbits of some components, $x_{l\,i,j}$, of the state variables $\mathbf{X}$ for the first 10,000 iterations are shown in Figs. 1.4.1(a)–(d).

It can be observed that the dynamic behaviors of the neighboring cells at the lattice, namely, (12, 11), (12, 10), (12, 12) and (13, 11), are quite different.

Now, construct an invertible transformation H, as follows:

$$H = \boldsymbol{B} \circ \tilde{H} : \mathbb{R}^{2\times M\times M} \to \mathbb{R}^{2\times M\times M}, \tag{1.4.4}$$

where

$$\tilde{H} = (\tilde{h}_1, \tilde{h}_2) : \mathbb{R}^{2\times M\times M} \to \mathbb{R}^{2\times M\times M} \tag{1.4.5}$$

and, for each fixed (i,j),

$$\boldsymbol{B} : \mathbb{R}^2 \to \mathbb{R}^2,$$

such that

$$\tilde{h}_1((x_{1\,i,j})_{21\times21}) = (\alpha_{i,j}^1)_{21\times21}(x_{1\,i,j})_{21\times21}, \tag{1.4.6}$$

$$\tilde{h}_2((x_{2\,i,j})_{21\times21}) = (\alpha_{i,j}^2)_{21\times21}(x_{2\,i,j})_{21\times21} \tag{1.4.7}$$

$$\mathbf{A}_1 = (\alpha_{i,j}^1)_{21\times21}, \tag{1.4.8}$$

$$\mathbf{A}_2 = (\alpha_{i,j}^2)_{21\times21}, \tag{1.4.9}$$

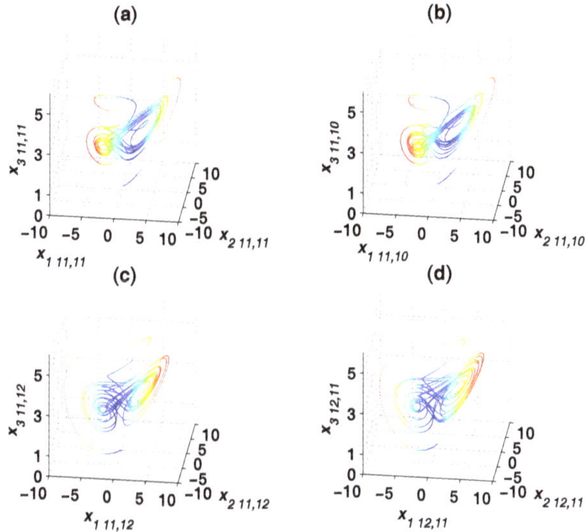

Fig. 1.4.1: Chaotic trajectories of some components of the state variables: (a) $x_{1\,12,11} - x_{2\,12,11} - x_{3\,12,11}$, (b) $x_{1\,12,10} - x_{2\,12,10} - x_{3\,12,10}$, (c) $x_{1\,12,12} - x_{2\,12,12} - x_{3\,12,12}$ and (d) $x_{1\,13,11} - x_{2\,13,11} - x_{3\,13,11}$.

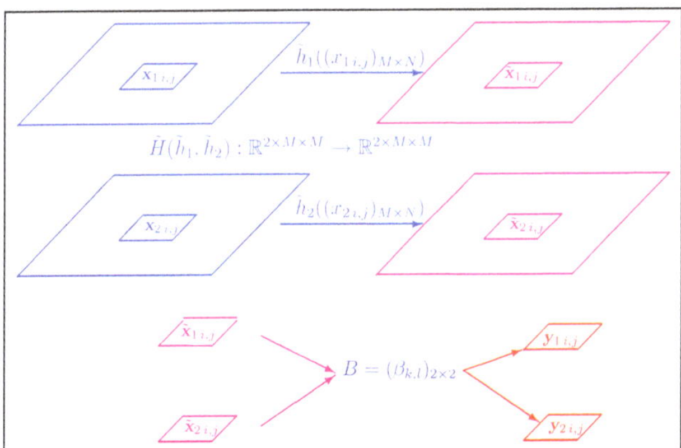

Fig. 1.4.2: Transformation $H = \boldsymbol{B} \circ \tilde{H} : \mathbb{R}^{2\times 21\times 21} \to \mathbb{R}^{2\times 21\times 21}$.

and

$$\boldsymbol{B} = \begin{bmatrix} 23 & 46 \\ 32 & 25 \end{bmatrix}, \tag{1.4.10}$$

are all invertible transformations (see Fig. 1.4.2).

By Theorem 1.1 and using the transformation H defined by (1.4.4)–(1.4.10), one can construct a GS system as follows:

$$\mathbf{Y}(k+1) = H((F_2(\mathbf{X}(k)) - Q(\mathbf{Y}, \mathbf{X}), \tag{1.4.11}$$

where

$$Q(\mathbf{Y}, \mathbf{X}) = \frac{1}{10}\mathbf{e}.$$

If the initial condition is selected as

$$\mathbf{Y}(0) = H(\mathbf{X}_2(0)) + (\delta_{l\,i,j})_{2\times 21\times 21}, \delta_{l\,i,j} \equiv 1, \tag{1.4.12}$$

then, the chaotic trajectories of the components $x_{k\,11,11}$ and $y_{k\,11,11}$ of the state variables $\mathbf{X}_2$ and $\mathbf{Y}$ for the first 10,000 iterations are obtained, as shown in Figs. 1.4.3(a) and (b).

Now, GS relationship between $\mathbf{X}_2$ and $\mathbf{Y}$ can be observed. Figures 1.4.3(c) and (d) show that, although the initial condition (1.4.12) has a perturbation, $\mathbf{X}_2$ and $\mathbf{Y}$ are rapidly reach GS, as the theory predicts.

The evolution of state variables: $t - x_{1\,11,11}$, $t - x_{2\,11,11}$, $t - x_{3\,11,11}$, $t - y_{1\,11,11}$, $t - y_{2\,11,11}$, and $t - y_{3\,11,11}$, are shown in Fig. 1.4.4. However, the GS between $x_{l\,i,j}$ and $y_{l\,i,j}$, $l = 1, 2$, cannot be seen there.

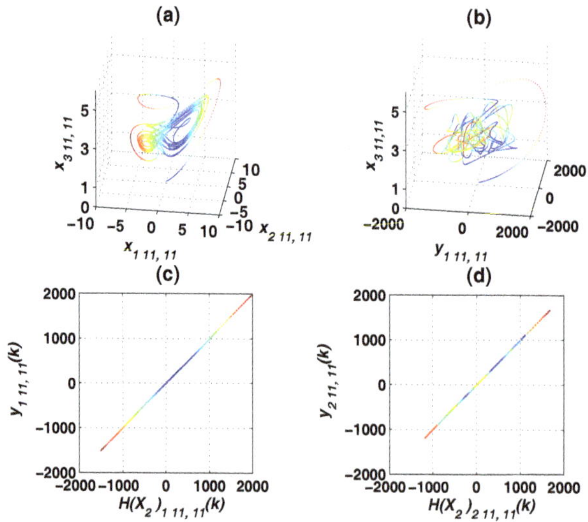

Fig. 1.4.3: Chaotic GS trajectories of some components of the state variables: (a) $x_{1\,11,11}$ - $x_{2\,11,11} - x_{3\,11,11}$, (b) $y_{1\,11,11} - y_{2\,11,11} - y_{3\,1,1}$, (c) $H(\mathbf{X}_2)_{1\,11,11}(k)$ and $y_{1\,11,11}(k)$ are in GS, (d) $H(\mathbf{X}_2)_{2\,11,11}(k)$ and $y_{2\,11,11}(k)$ are in GS.

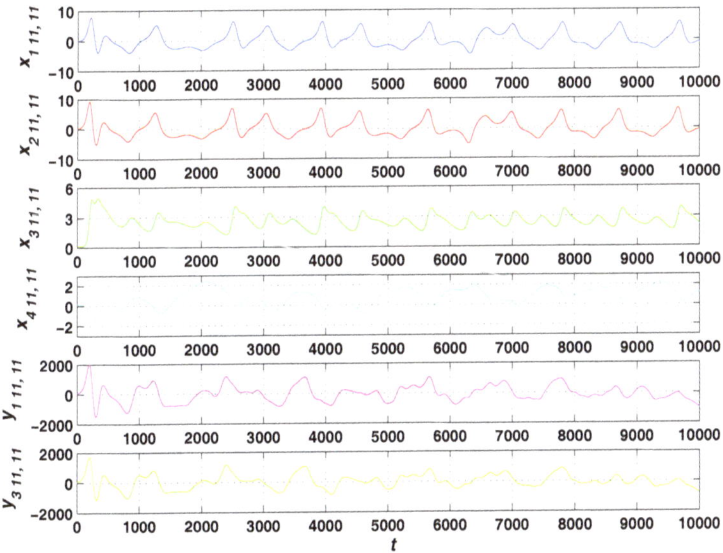

Fig. 1.4.4: The evolution of state variables: $t - x_{1\,11,11}$, $t - x_{2\,11,11}$, $t - x_{3\,11,11}$, $t - x_{4\,11,11}$, $t - y_{1\,11,11}$, and $t - y_{2\,11,11}$.

Fig. 1.4.5: Pseudo-color code for the components of the solution of the Chen CNN, where $G_{i,j}$ stands for $x_{l\,i,j}$ or $y_{l\,i,j}$.

To display the complex GS patterns of the discrete Chen CNN, choose a pseudo-color code, as shown in Fig. 1.4.5. The evolution of the chaotic patterns of the state variables $x_{k\,i,j}$, $y_{k\,i,j}$ for the first 10,000 iterations are shown in Fig. 1.4.6.

It can be observed that the randomly perturbed initial patterns are evolving irregularly but the GS relationship between $x_{l\,i,j}$ and $y_{l\,i,j}$, $l = 1, 2$ cannot be seen.

The three-dimensional views of the evolution of the Chen CNN at different iterative steps k are shown in Fig. 1.4.7, in which chaotic waves can be seen clearly.

In summary, the discrete Chen CNN has extremely complex dynamic behaviors.

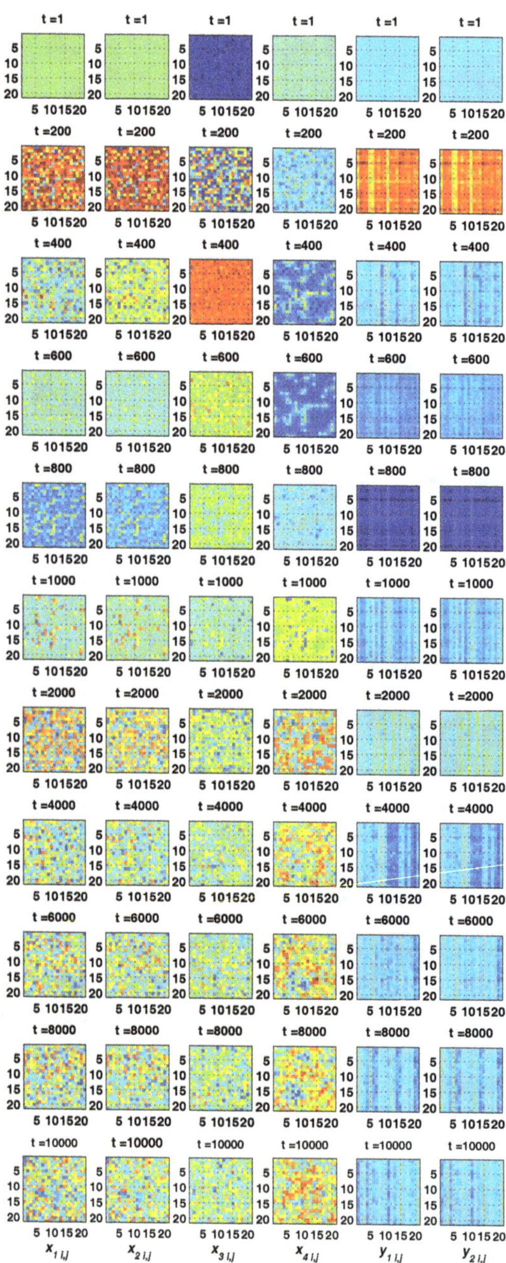

Fig. 1.4.6: Evolution of patterns in the discrete Chen CNN on the first 10,000 iterations.

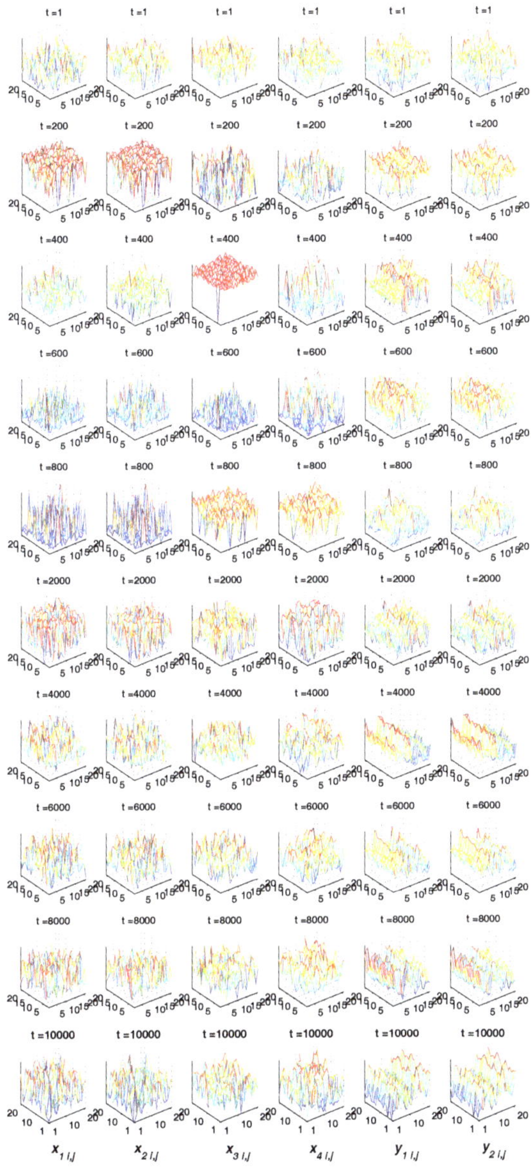

Fig. 1.4.7: Three-dimensional views of the discrete Chen CNN spiral waves at different iteration steps k. The vertical axes represent the state variables $x_{1\,i,j}$, $x_{2\,i,j}$, $x_{3\,i,j}$, $x_{4\,i,j}$, $y_{1\,i,j}, y_{2\,i,j}$, while the horizontal axes are the plane coordinates (i, j).

1.4.2 Application of GS Theorem to Non-autonomous CDADS

In this subsection, an application of Theorem 1.3 to the GS of CDADS is discussed. Based on the Chen CNN introduced in Subsection 1.4.1 above, a non-autonomous discrete Chen CNN is introduced as follows:

$$\begin{cases} x_{1\,i,j}(k+1) = x_{1\,i,j}(k) + \epsilon a[x_{2\,i,j}(k) - x_{1\,i,j}(k) + \sin(k)] \\ x_{2\,i,j}(k+1) = x_{2\,i,j}(k) + \epsilon[4x_{1,i,j}(k) + cx_{2\,i,j}(k) \\ \qquad\qquad -10x_{1\,i,j}(k)x_{3\,i,j}(k) + 4x_{4\,i,j}(k)] \\ x_{3\,i,j}(k+1) = x_{3\,i,j}(k) + \epsilon[x_{2\,i,j}^2(k) - bx_{3\,i,j}(k)] \\ x_{4\,i,j}(k+1) = x_{4\,i,j}(k) - \epsilon dx_{1\,i,j}(k) \\ \qquad\qquad D_2[x_{3i+1,j} + x_{3i-1,j} \\ \qquad\qquad +x_{3i,j+1} + x_{3i,j-1} - 4x_{4i,j}], \\ \qquad\qquad i,j = 1,2,\cdots,21, \end{cases} \tag{1.4.13}$$

where

$$a = 35, b = 3, c = 21, d = 2, \epsilon = 0.001, D = 0.01.$$

Written in a compact form, it is

$$\mathbf{X}(k+1) = F(\mathbf{X}(k)). \tag{1.4.14}$$

Now, select the same invertible transformation H defined by (1.4.4)-(1.4.10):

$$H = \mathbf{B} \circ \tilde{H} : \mathbb{R}^{2\times M\times M} \to \mathbb{R}^{2\times M\times M}.$$

Using Theorem 1.3, one can construct a GS system as follows:

$$\mathbf{Y}(k+1) = H(F_2(\mathbf{X}(k))) - Q(\mathbf{Y}, \mathbf{X}), \tag{1.4.15}$$

where

$$Q(\mathbf{Y}, \mathbf{X}) = \frac{1}{10}\mathbf{e}. \tag{1.4.16}$$

Next, select the same initial conditions (1.4.3) and (1.4.12) for $\mathbf{X}(0)$ and $\mathbf{Y}(0)$. The chaotic orbits of some components $x_{l\,i,j}$ of the state variables $\mathbf{X}$ for the first 10,000 iterations are shown in Figs. 1.4.8 (a)–(d). It can be observed that the dynamic behaviors of the neighboring cells at the lattice: (12, 11), (12, 10), (12, 12) and (13, 11), are quite different.

The chaotic trajectories of the components $x_{k\,11,11}$ and $y_{k\,11,11}$ of the state variables $\mathbf{X}$ and $\mathbf{Y}$ for the first 10,000 iterations are shown in Figs. 1.4.9(a) and (b).

Figures 1.4.9(c) and (d) show that, although the initial condition (1.4.12) has a perturbation, $\boldsymbol{X}_2$ and $\boldsymbol{Y}$ are rapidly reach GS, as the theory predicts.

The evolution of state variables: $t - x_{1\,11,11}$, $t - x_{2\,11,11}$, $t - x_{3\,11,11}$, $t - y_{1\,11,11}$, $t - y_{2\,11,11}$, and $t - y_{3\,11,11}$ are shown in Fig. 1.4.10. However, the GS relationships between $x_{l\,i,j}$ and $y_{l\,i,j}, l = 1, 2$ cannot be seen.

The evolution of chaotic patterns of the state variables $x_{k\,i,j}$, $y_{k\,i,j}$ over the first 10,000 iterations are shown in Fig. 1.4.11. It can be observed that the randomly perturbed initial patterns are evolving irregularly, but the GS relationships between $x_{l\,i,j}$ and $y_{l\,i,j}, l = 1, 2$ cannot be seen.

The three-dimensional views of the evolution of the non-autonomous Chen CNN at different iterative steps k are shown in Fig. 1.4.12, in which chaotic waves can be seen clearly.

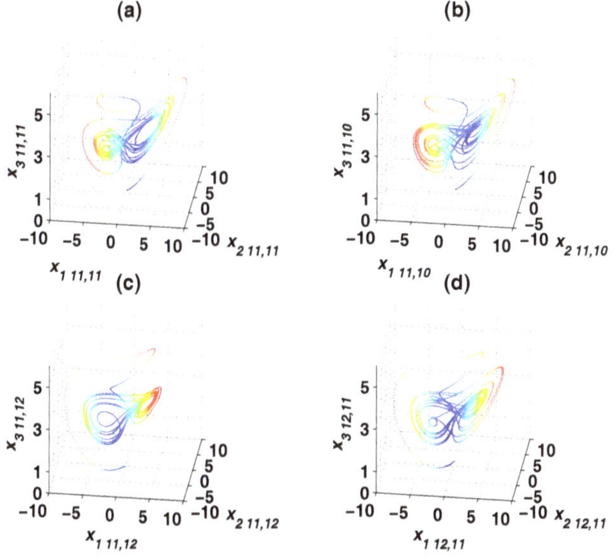

Fig. 1.4.8: Chaotic trajectories of some components of the state variables: (a) $x_{1\,12,11} - x_{2\,12,11} - x_{3\,12,11}$, (b) $x_{1\,12,10} - x_{2\,12,10} - x_{3\,12,10}$, (c) $x_{1\,12,12} - x_{2\,12,12} - x_{3\,12,12}$ and (d) $x_{1\,13,11} - x_{2\,13,11} - x_{3\,13,11}$.

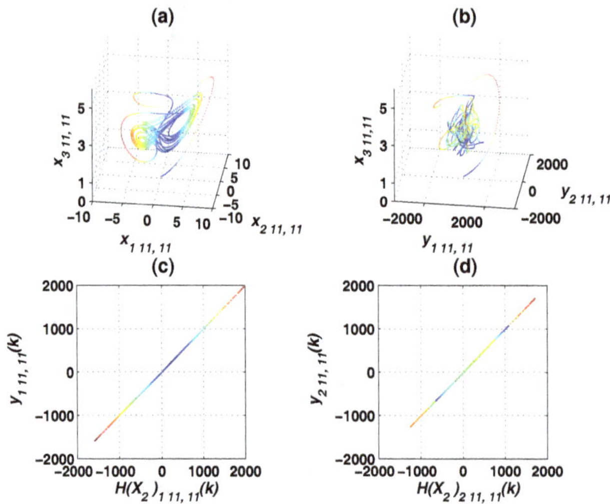

Fig. 1.4.9: Chaotic GS trajectories of some components of the state variables: (a) $x_{1\,11,11}$ - $x_{2\,11,11} - x_{3\,11,11}$, (b) $y_{1\,11,11} - y_{2\,11,11} - y_{3\,1,1}$, (c) $H(\mathbf{X}_2)_{1\,11,11}(k)$ and $y_{1\,11,11}(k)$ are in GS, (d) $H(\mathbf{X}_2)_{2\,11,11}(k)$ and $y_{2\,11,11}(k)$ are in GS.

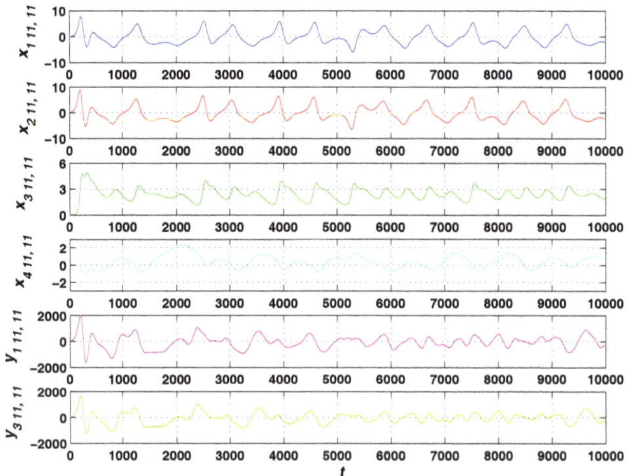

Fig. 1.4.10: The evolution of state variables: $t - x_{1\,11,11}$, $t - x_{2\,11,11}$, $t - x_{3\,11,11}$, $t - x_{4\,11,11}$, $t - y_{1\,11,11}$, and $t - y_{2\,11,11}$.

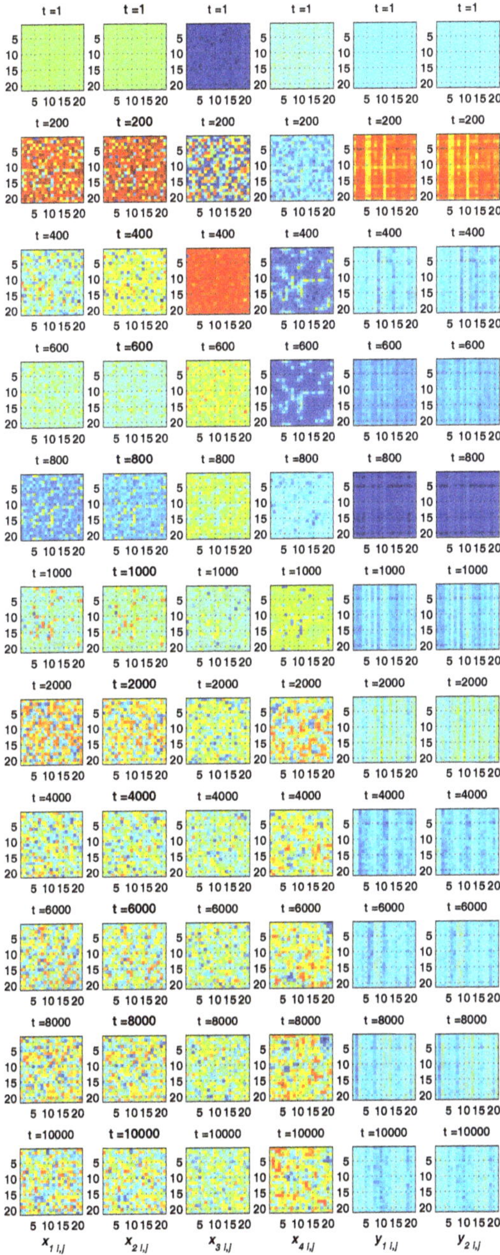

Fig. 1.4.11: Evolution of patterns of the discrete non-autonomous Chen CNN over the first 10,000 iterations.

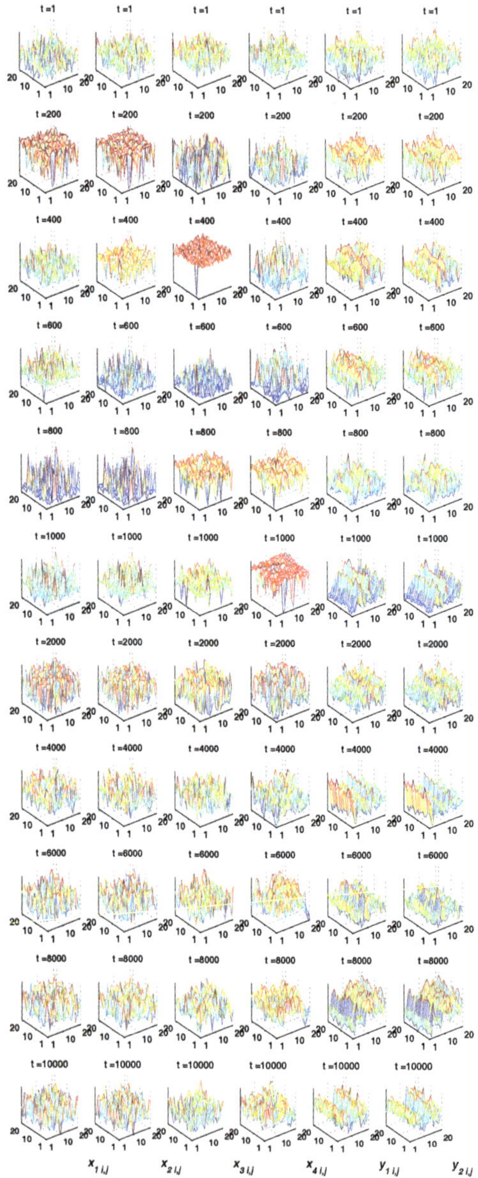

Fig. 1.4.12: Three-dimensional, views of the discrete Chen CNN spiral waves at different iteration steps k. The vertical axes represent the state variables $x_{1\,i,j}$, $x_{2\,i,j}$, $x_{3\,i,j}$, $x_{4\,i,j}$, $y_{1\,i,j}$, $y_{2\,i,j}$, while the horizonal axes are the plane coordinates (i, j).

In summary, comparing Fig. 1.4.1, Fig. 1.4.3, Fig. 1.4.4, Fig. 1.4.6 and Fig. 1.4.7 with Figs. 1.4.8–1.4.12 reveals that a small non-autonomous perturbation term $3.5\sin(k) \times 10^{-2}$ in the first equation of the Chen CNN (1.4.13) may make the dynamic behaviors of the Chen CNN become quite different.

1.4.3 *Application of GS Theorem to Non-autonomous Bidirectional CDADS*

In this subsection, an application of Theorem 1.4 is presented. Firstly, a coupled non-autonomous discrete bidirectional Lorenz CNN is constructed, whose prototype is the 3D discrete Lorenz equation [Sprot (2003)].

Specifically, a continuously differentiable and invertible transformation H is constructed as follows:

$$H = \boldsymbol{B} \circ \tilde{H} : \mathbb{R}^{3\times M\times M} \to \mathbb{R}^{3\times M\times M} \tag{1.4.17}$$

where

$$\tilde{H} = (\tilde{h}_1, \tilde{h}_2, \tilde{h}_3) : \mathbb{R}^{3\times M\times M} \to \mathbb{R}^{3\times M\times M}, \tag{1.4.18}$$

and

$$\boldsymbol{B} = (\beta_{k,l})_{3\times 3} : \mathbb{R}^3 \to \mathbb{R}^3, \tag{1.4.19}$$

such that

$$\begin{aligned}\tilde{h}_1((x_{1\,i,j})_{21\times 21}) &= (\alpha^1_{i,j})_{21\times 21}(x_{1\,i,j})_{21\times 21}\\ &= (\tilde{x}_{1\,i,j})_{21\times 21},\end{aligned} \tag{1.4.20}$$

$$\begin{aligned}\tilde{h}_2((x_{2\,i,j})_{21\times 21}) &= (\alpha^2_{i,j})_{21\times 21}(x_{2\,i,j})_{21\times 21}\\ &= (\tilde{x}_{2\,i,j})_{21\times 21},\end{aligned} \tag{1.4.21}$$

$$\begin{aligned}\tilde{h}_3((x_{3\,i,j})_{21\times 21}) &= (\alpha^2_{i,j})_{21\times 21}(x_{3\,i,j})_{21\times 21}\\ &= (\tilde{x}_{3\,i,j})_{21\times 21},\end{aligned} \tag{1.4.22}$$

and, for any triple $(\tilde{x}_{1\,i,j}, \tilde{x}_{2\,i,j}, \tilde{x}_{3\,i,j})$,

$$\boldsymbol{B}(\tilde{x}_{1\,i,j}, \tilde{x}_{2\,i,j}, \tilde{x}_{3\,i,j}) = (\beta_{k,l})_{3\times 3}[\tilde{x}_{1\,i,j}, \tilde{x}_{2\,i,j}, \tilde{x}_{3\,i,j}]^{\mathrm{T}}$$

where

$$\begin{cases}\boldsymbol{A}_1 = (\alpha^1_{i,j})_{21\times 21}\\ \boldsymbol{A}_2 = (\alpha^2_{i,j})_{21\times 21}\\ \boldsymbol{A}_3 = (\alpha^3_{i,j})_{21\times 21}\end{cases} \tag{1.4.23}$$

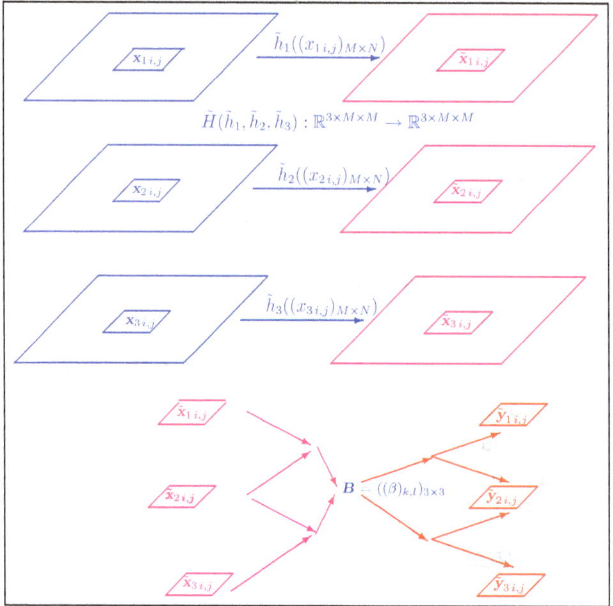

Fig. 1.4.13: Transformation $H = \boldsymbol{B} \circ \tilde{H} : \mathbb{R}^{3\times21\times21} \to \mathbb{R}^{3\times21\times21}$.

and

$$\boldsymbol{B} = \begin{bmatrix} 0.001 & -0.039 & 0.037 \\ 0.042 & -0.062 & -0.040 \\ -0.014 & -0.062 & 0.008 \end{bmatrix} \tag{1.4.24}$$

are all invertible transformations (see Fig. 1.4.13).

Using Theorem 1.4, one can construct a non-autonomous bidirectional Lorenz CNN as follows:

$$\begin{cases} x_{1\,i,j}(k+1) = x_{1\,i,j}(k)x_{2\,i,j}(k) - x_{3\,i,j}(k) + k_0\sin(k) \\ \qquad\qquad\qquad +k_0\sin(\pi y_{1\,i,j}(k)y_{2\,i,j}(k)y_{3\,i,j}(k)) \\ x_{2\,i,j}(k+1) = x_{1\,i,j}(k) \\ x_{3\,i,j}(k+1) = x_{2\,i,j}(k) + D_2[x_{3i+1,j} + x_{3i-1,j} \\ \qquad\qquad\qquad +x_{3i,j+1} + x_{3i,j-1} - 4x_{4i,j}], \\ \qquad\qquad\qquad i,j = 1,2,\cdots,21, \end{cases} \tag{1.4.25}$$

where $k_0 = 1e-6, D = 0.01$. Written in a compact form, it is

$$\boldsymbol{X}(k+1) = F(\boldsymbol{X}(k)). \tag{1.4.26}$$

The second part of the Lorenz CNN has the following form:

$$\boldsymbol{Y}(k+1) = H((F(\mathbf{X}(k)) - Q(\mathbf{Y},\mathbf{X}), \tag{1.4.27}$$

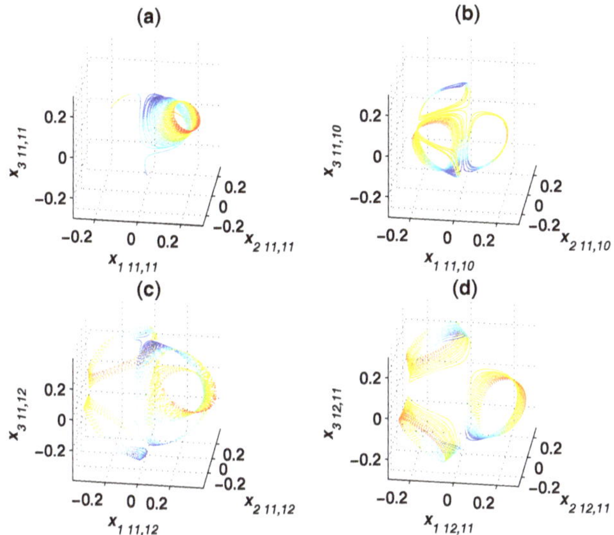

Fig. 1.4.14: Chaotic trajectories of some components of the state variables: (a) $x_{1\,12,11} - x_{2\,12,11} - x_{3\,12,11}$, (b) $x_{1\,12,10} - x_{2\,12,10} - x_{3\,12,10}$, (c) $x_{1\,12,12} - x_{2\,12,12} - x_{3\,12,12}$ and (d) $x_{1\,13,11} - x_{2\,13,11} - x_{3\,13,11}$.

where H is defined by (1.4.17)–(1.4.24), and

$$Q(\boldsymbol{Y}, \boldsymbol{X}) = \frac{1}{10}\mathbf{e}. \tag{1.4.28}$$

Now, select the following initial conditions:

$$(x_{l\,i,j}(0))_{21\times 21} = 0.1(1 + rand(21, 21)), l = 1, 2, 3, \tag{1.4.29}$$

$$\boldsymbol{Y}(0) = H(\boldsymbol{X}(0)) + (\delta_{l\,i,j})_{3\times 21\times 21}, \delta_{l\,i,j} \equiv 1,$$

where $rand(21, 21)$ is a Matlab command.

The chaotic orbits of some components $x_{l\,i,j}$ of the state variables $\mathbf{X}$ over the first 10,000 iterations are shown in Figs. 1.4.14(a)– (d). It can be observed that the dynamic behaviors of the neighboring cells at the lattice: (11, 11), (11, 10), (11, 12) and (12, 11), are completely different.

The chaotic trajectories of the components $x_{k\,11,11}$ and $y_{k\,11,11}$ of the state variables $\boldsymbol{X}$ and $\boldsymbol{Y}$ over the first 10,000 iterations are shown in Figs. 1.4.15(a) and (b).

Figures 1.4.15(c) and 1.4.15(d) show that, although the initial condition (1.4.30) has a perturbation, $\boldsymbol{X}$ and $\boldsymbol{Y}$ rapidly reach GS, as the theory predicts.

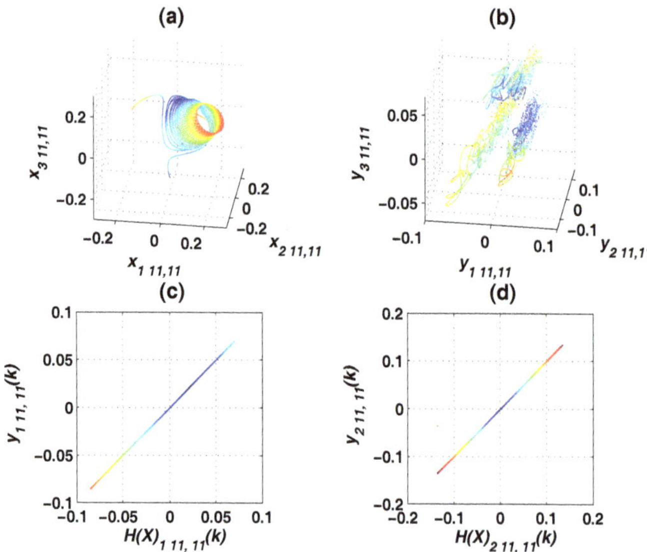

Fig. 1.4.15: Chaotic GS trajectories of some component of the state variables: (a) $x_{1\,11,11} - x_{2\,11,11} - x_{3\,11,11}$, (b) $y_{1\,11,11} - y_{2\,11,11} - y_{3\,1,1}$, (c) $H(\boldsymbol{X})_{1\,11,11}(k)$ and $y_{1\,11,11}(k)$ are in GS, (d) $H(\boldsymbol{X})_{2\,11,11}(k)$ and $y_{2\,11,11}(k)$ are in GS.

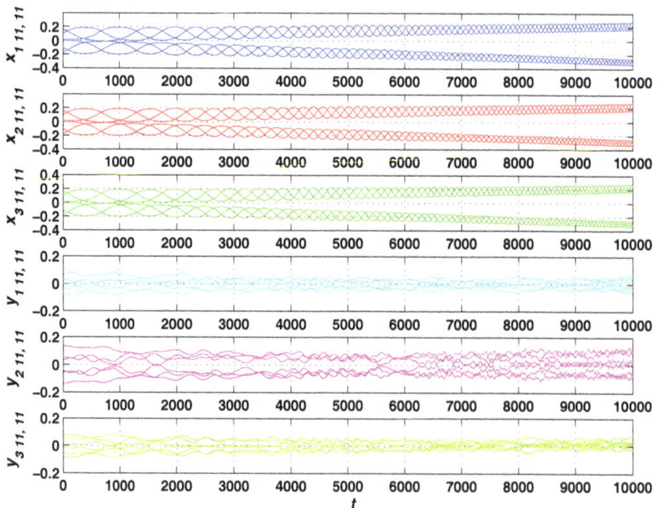

Fig. 1.4.16: The evolution of state variables: $t - x_{1\,11,11}$, $t - x_{2\,11,11}$, $t - x_{3\,11,11}$, $t - x_{4\,11,11}$, $t - y_{1\,11,11}$, and $t - y_{2\,11,11}$.

The evolution of state variables: $t - x_{1\,11,11}$, $t - x_{2\,11,11}$, $t - x_{3\,11,11}$, $t - y_{1\,11,11}$, $t - y_{2\,11,11}$, and $t - y_{3\,11,11}$ are shown in Fig. 1.4.16. However, the GS relationships between $x_{l\,i,j}$ and $y_{l\,i,j}, l = 1, 2, 3$ cannot be seen.

The evolution of the chaotic patterns of the state variables $x_{k\,i,j}$, $y_{k\,i,j}$ over the first 10,000 iterations are shown in Fig. 1.4.17. It can be observed that the randomly perturbed initial patterns ($k = 1$) are evolving irregularly but the GS relationships between $x_{l\,i,j}$ and $y_{l\,i,j}, l = 1, 2$ cannot be seen.

The three-dimensional views of the evolution of the Lorenz CNN at different iterative steps k are shown in Fig. 1.4.18, in which chaotic waves can be seen clearly.

In summary, the non-autonomous discrete bidirectional Lorenz CNN has extremely complex dynamic behaviors. In particular, the dynamic behaviors of neighboring cells of the CNN may be quite different.

1.4.4 *Application of GS Theorem to Bidirectional CCADS*

In this subsection, an application of Theorem 1.7 to bidirectional CCADS is discussed.

Based on the third-order Chua circuit [Chua *et al.* (1994)]), and a smooth Chua CNN with $3 \times 21 \times 21$ arrays and one-port [Min *et al.* (2000)], a bidirectional Chua CNN is introduced.

The GS transformation H of the Chua CNN is the same as that defined by (1.4.17)–(1.4.24).

The first part of the Chua CNN has the following form:

$$\begin{cases} \dot{x}_{1i,j} = k\alpha x_{2i,j} - x_{1i,j} - bx_{1i,j} - \dfrac{(a-b)}{\pi}\arctan(ux_{1i,j}) \\ \dot{x}_{2i,j} = x_{1i,j} - x_{2i,j} + x_{3i,j} + \epsilon\sin(\pi y_{1i,j}y_{2i,j}y_{3i,j}) \\ \dot{x}_{3i,j} = -\beta x_{2i,j} - \gamma x_{3i,j} + D[x_{3i+1,j} + x_{3i-1,j} \\ \qquad +x_{3i,j+1} + x_{3i,j-1} - 4x_{4i,j}], \\ \qquad i, j = 1, 2, \ldots, 21, \end{cases} \tag{1.4.30}$$

where $\alpha = 10, \beta = 15, \gamma = 0.3$, $a = -1.2, b = -0.5, k = 1, u = 5, D = 0.01, \epsilon = 1e-6$.

In a compact form, system (1.4.30) can be written as

$$\dot{\mathbf{X}} = F(\mathbf{X}, \mathbf{Y}). \tag{1.4.31}$$

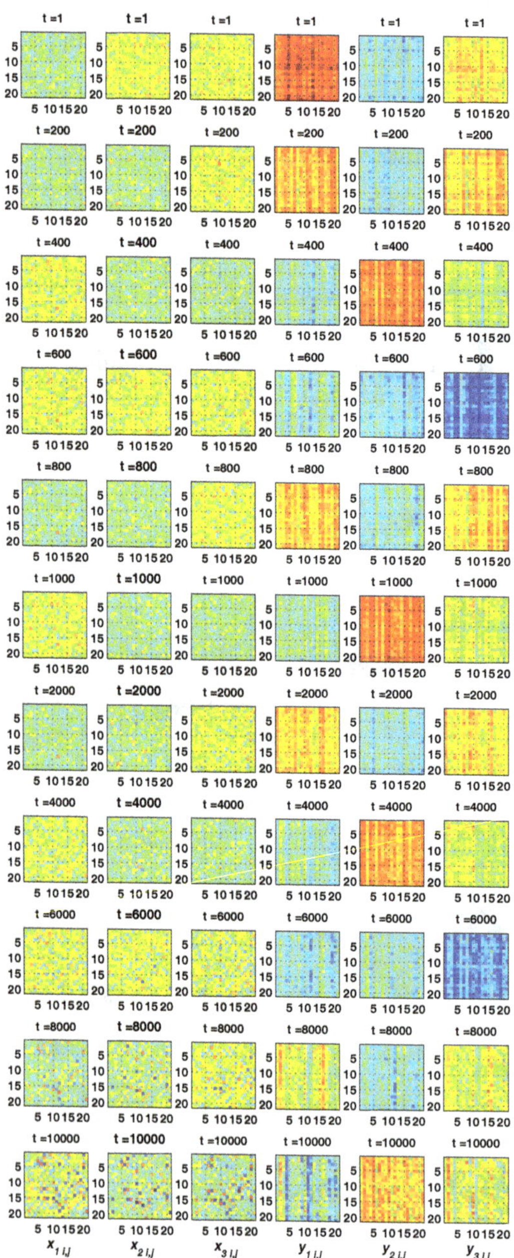

Fig. 1.4.17: Evolution of patterns of the discrete non-autonomous bidirectional Lorenz CNN over the first 10,000 iterations.

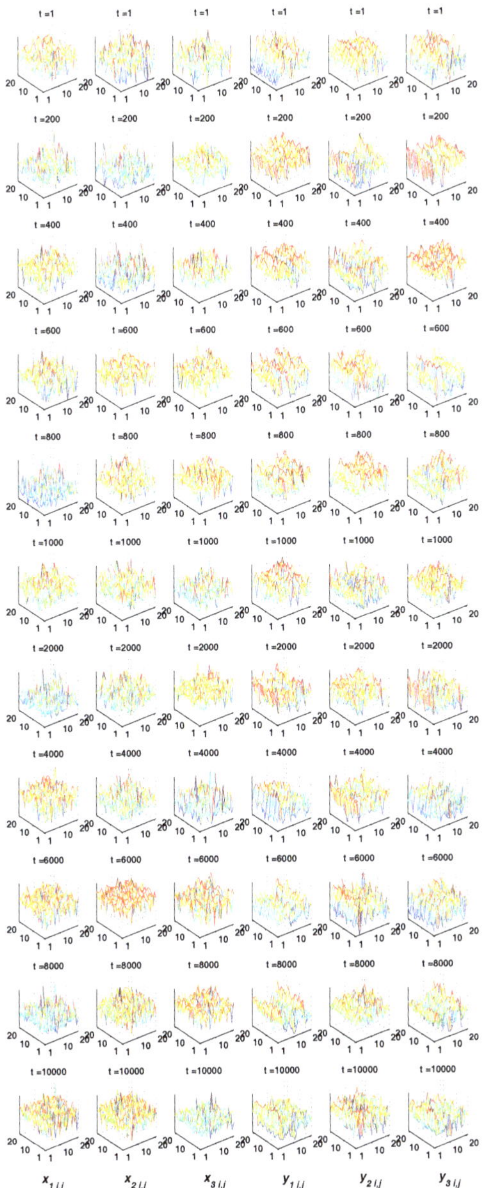

Fig. 1.4.18: The three-dimensional views of discrete non-autonomous bidirectional Lorenz CNN spiral waves at different iteration steps k. The vertical axes represent the state variables $x_{1\,i,j}$, $x_{2\,i,j}$, $x_{3\,i,j}$, $y_{1\,i,j}$, $y_{2\,i,j}$, $y_{3\,i,j}$, while the horizontal axes are the plane coordinates (i, j).

Let

$$\dot{\mathbf{Y}} = G(\mathbf{X}, \mathbf{Y}), \tag{1.4.32}$$

where

$$\begin{aligned} G_{l\,i,j}(\mathbf{Y}, \mathbf{X}) = & \sum_{l'=1}^{3} \sum_{i'=1}^{21} \sum_{j'=1}^{21} \frac{\partial h_{l\,i,j}(\mathbf{X})}{\partial x_{l'\,i',j'}} f_{l'\,i',j'}(\mathbf{X}, \mathbf{Y}) \\ & - q_{l\,i,j}(\mathbf{Y}, \mathbf{X}), \end{aligned} \tag{1.4.33}$$

$$q_{l\,i,j}(\mathbf{Y}, \mathbf{X}) = h_{l\,i,j}(\mathbf{X}) - y_{l\,i,j}. \tag{1.4.34}$$

By Theorem 1.5, systems (1.4.31) and (1.4.32) are in GS with respect to the transformation H defined by (1.4.17)–(1.4.24) (see Fig. 1.4.13).

Now, select the following initial conditions:

$$\begin{aligned} (x_{l\,i,j}(0))_{21\times 21} = & \mathbf{X}_0(l) + 0.02(rand(21, 21) - 0.5), \\ & l = 1, 2, 3, \end{aligned}$$

$$\begin{aligned} \mathbf{Y}(0) = & H(\mathbf{X}(0)) \\ & + 0.02(rand(3, 21, 21) - 0.5) \end{aligned} \tag{1.4.35}$$

where $\mathbf{X}_0 = [0.41379, 0.027763, -0.012759]$ and $rand(3, 21, 21)$ is a Matlab command.

The chaotic trajectories of some components $x_{l\,i,j}$ of the state variables $\mathbf{X}$ over the time interval [0, 50] are shown in Fig. 1.4.19(a)–(d). It can be observed that the dynamic behaviors of the neighboring cells at the lattice: (12,11), (12, 10), (12, 12) and (13, 11), are quite different.

The chaotic trajectories of some components $x_{k\,11,11}$ and $y_{k\,11,11}$ of the state variables $\mathbf{X}$ and $\mathbf{Y}$ over the time interval [0, 50] are shown in Figs. 1.4.20(a) and (b).

Figures 1.4.20(c) and 1.4.20(d) show that, although the initial condition (1.4.35) has a perturbation, $\mathbf{X}$ and $\mathbf{Y}$ rapidly reach GS, as the theory predicts.

The evolution of state variables $t - x_{1\,11,11}$, $t - x_{2\,11,11}$, $t - x_{3\,11,11}$, $t - y_{1\,11,11}$, $t - y_{2\,11,11}$, and $t - y_{3\,11,11}$ is shown in Fig. 1.4.21.

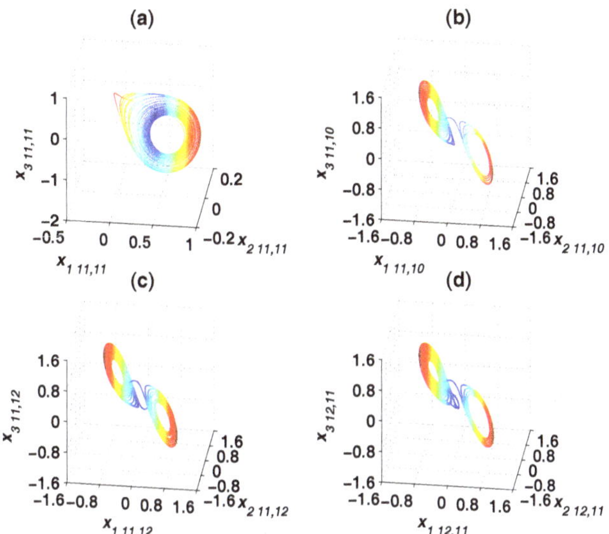

Fig. 1.4.19: Chaotic trajectories of some components of the state variables: (a) $x_{1\,12,11} - x_{2\,12,11} - x_{3\,12,11}$, (b) $x_{1\,12,10} - x_{2\,12,10} - x_{3\,12,10}$, (c) $x_{1\,12,12} - x_{2\,12,12} - x_{3\,12,12}$, and (d) $x_{1\,13,11} - x_{2\,13,11} - x_{3\,13,11}$.

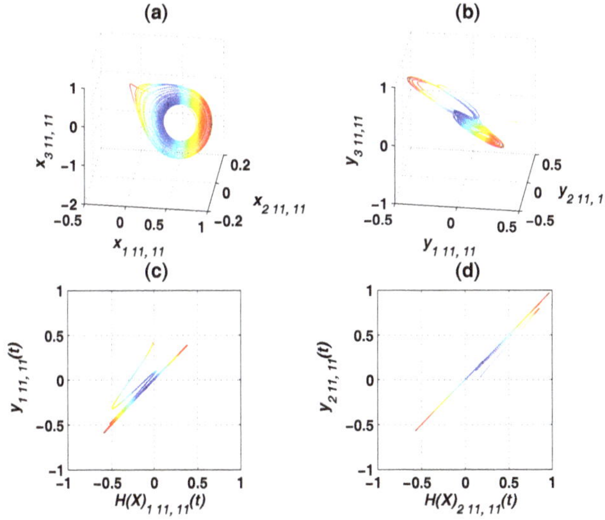

Fig. 1.4.20: Chaotic GS trajectories of some components of the state variables: (a) $x_{1\,11,11}- x_{2\,11,11}- x_{3\,11,11}$, (b) $y_{1\,11,11}- y_{2\,11,11}- y_{3\,1,1}$, (c) $H(\mathbf{X})_{1\,11,11}(k)$ and $y_{1\,11,11}(k)$ are in GS, (d) $H(\mathbf{X})_{2\,11,11}(k)$ and $y_{2\,11,11}(k)$ are in GS.

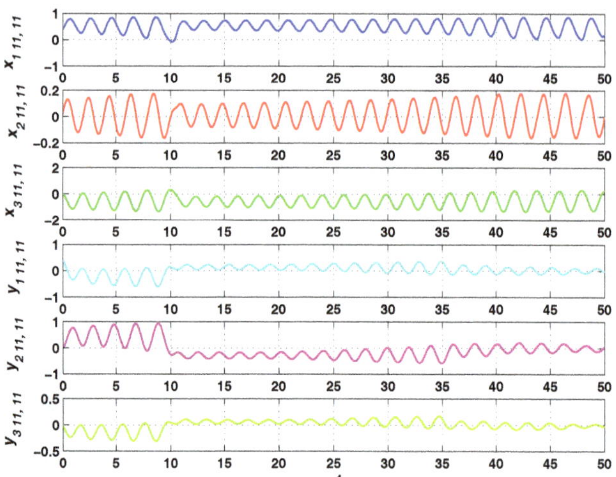

Fig. 1.4.21: The evolution of state variables: $t - x_{1\,11,11}$, $t - x_{2\,11,11}$, $t-$ $x_{3\,11,11}$, $t - x_{4\,11,11}$, $t - y_{1\,11,11}$, and $t - y_{2\,11,11}$.

The evolution of the chaotic patterns of the state variables $x_{k\,i,j}$, $y_{k\,i,j}$ over the time interval [0, 50] are shown in Fig. 1.4.22. It can be observed that the randomly perturbed initial patterns ($t = 0$) are evolving irregularly but the GS relationships between $x_{l\,i,j}$ and $y_{l\,i,j}, l = 1, 2, 3$ cannot be seen.

The three-dimensional views of the evolution of the Chua CNN at different times are shown in Fig. 1.4.23, in which chaotic waves can be seen clearly. It can be observed that the irregular chaotic waves shown in the first three columns in Fig 1.4.23 have been transformed to wall-shaped chaotic wave forms shown in the last three columns in Fig 1.4.23.

In summary, the bidirectional Chua CNN has extremely complex dynamic behaviors.

1.4.5 *Application of GS Theorem to Non-autonomous Bidirectional CCADS*

In this subsection, an application of Theorem 1.9 to non-autonomous bidirectional CCADS is discussed.

Based on the third-order Chen system [Chen and Uets (1999)], a non-autonomous bidirectional Chen CNN is introduced, as follows:

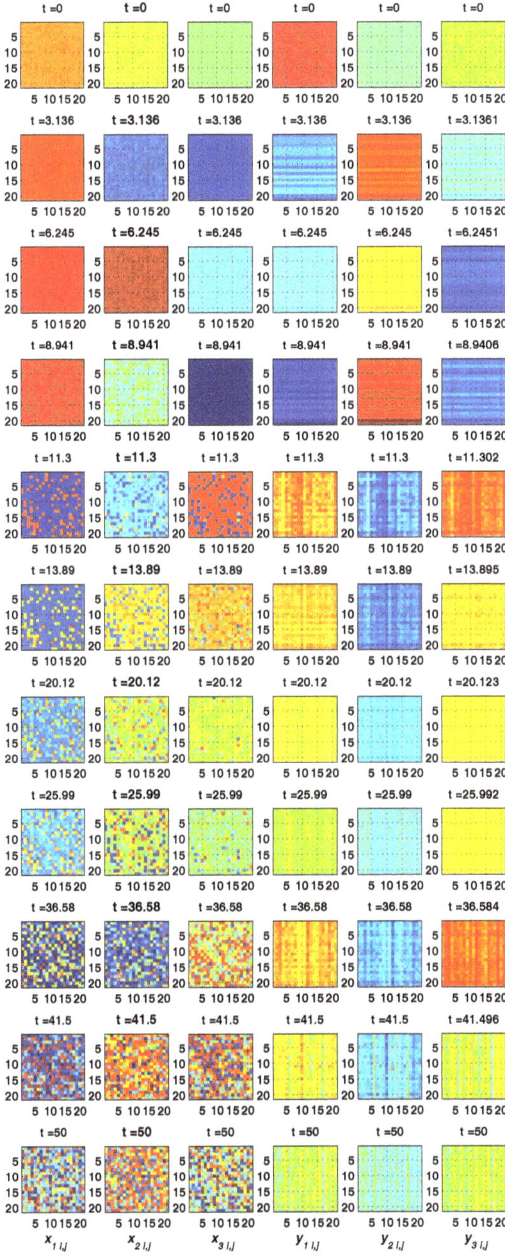

Fig. 1.4.22: Evolution of patterns of the bidirectional Chua CNN over the first 10,000 iterations.

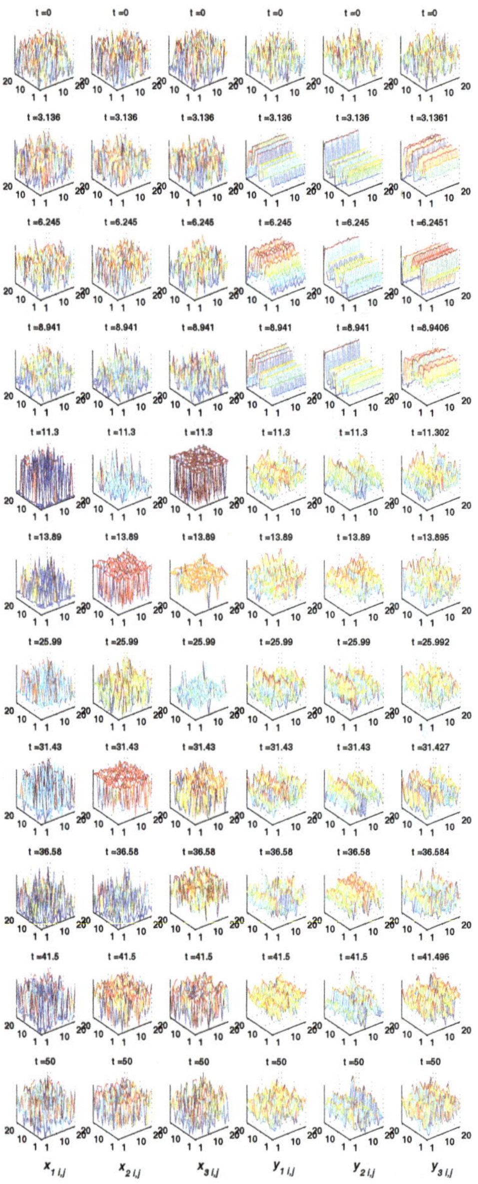

Fig. 1.4.23: The three-dimensional views of the discrete bidirectional Chua CNN spiral waves at different iteration steps k. The vertical axes represent the state variables, $x_{1\,i,j}$, $x_{2\,i,j}$, $x_{3\,i,j}$, $y_{1\,i,j}, y_{2\,i,j}$, while the horizontal axes are the spatial coordinates (i, j).

$$\begin{cases} \dot{x}_{1i,j} = a(x_{2i,j} - x_{1i,j}) + k_0 \cos(\pi t) \\ \dot{x}_{2i,j} = (c-a)x_{1i,j} - x_{1i,j}x_{3i,j} + cx_{2i,j} + D_1 \sin(\pi y_{1i,j} y_{2i,j} y_{3i,j}) \\ \dot{x}_{3i,j} = x_{1i,j}x_{2i,j} - bx_{3i,j} + D_2[x_{3i+1,j} + x_{3i-1,j} \\ \qquad +x_{3i,j+1} + x_{3i,j-1} - 4x_{4i,j}], \\ \qquad i,j = 1,2,\ldots,21, \end{cases} \tag{1.4.36}$$

where $a = 35, b = 3, c = 28, k_0 = 0.1, D_i = 5 \times 10^{-4}$.

In a compact form, the CCADS (1.4.36) can be written as

$$\dot{\mathbf{X}} = F(\mathbf{X}, \boldsymbol{Y}, t). \tag{1.4.37}$$

Now, let

$$\dot{\mathbf{Y}} = G(\mathbf{X}, \boldsymbol{Y}, t), \tag{1.4.38}$$

where

$$\begin{aligned} G_{l\,i,j}(\boldsymbol{Y}, \mathbf{X}, t) = &\sum_{l'=1}^{3}\sum_{i'=1}^{21}\sum_{j'=1}^{21} \frac{\partial h_{l\,i,j}(\mathbf{X})}{\partial x_{l'\,i',j'}} f_{l'\,i',j'}(\mathbf{X}, \boldsymbol{Y}) \\ &-q_{l\,i,j}(\boldsymbol{Y}, \mathbf{X}), \end{aligned} \tag{1.4.39}$$

$$\begin{aligned} H(\mathbf{X}) &= (h_{l\,i,j}(\mathbf{X}))_{3\times 21\times 21} \\ &= \left(\sum_{h=1}^{3} \beta_{l,h} \sum_{k=1}^{21}\sum_{m=1}^{21} \alpha_{h\,i,k} x_{h\,m,j}\right)_{3\times 21\times 21}. \end{aligned} \tag{1.4.40}$$

Here,

$$\mathbf{A}_l = (\alpha_{l\,i,k})_{l\times 21\times 21}, \tag{1.4.41}$$

$$\boldsymbol{B} = (\beta_{i,k})_{l\times 21\times 21} \tag{1.4.42}$$

$$l = 1,2,3,$$

are invertible matrices, and

$$q_{l\,i,j}(\boldsymbol{Y}, \mathbf{X}) = h_{l\,i,j}(\mathbf{X}) - y_{l\,i,j}. \tag{1.4.43}$$

By Theorem 1.9, systems (1.4.37) and (1.4.38) are in GS with respect to the transformation H defined by (1.4.40)–(1.4.43) (see Fig. 1.4.13).

Next, select the following initial conditions:

$$\begin{aligned} (x_{l\,i,j}(0))_{21\times 21} &= \mathbf{X}_0(l) + 0.02(rand(21,21) - 0.5), \\ l &= 1,2,3, \\ \boldsymbol{Y}(0) &= \mathbf{X}(0) + 0.02(rand(21,21) - 0.5), \end{aligned} \tag{1.4.44}$$

where $\boldsymbol{X}_0 = [1.1737\ 2.8428\ 19.103]$, $rand(21,21)$ is a Matlab command, which returns an 21×21 matrix containing pseudo-random values drawn from a uniform distribution on the unit interval.

The chaotic orbits of some components $x_{l\,i,j}$ of the state variables $\boldsymbol{X}$ over the time interval [0, 5] are shown in Fig. 1.4.24(a)–(d).

The chaotic trajectories of the components $x_{k\,11,11}$ and $y_{k\,11,11}$ of the state variables $\boldsymbol{X}$ and $\boldsymbol{Y}$ over the the time interval [0, 5] are shown in Figs. 1.4.25(a) and 1.4.25(b).

Figures 1.4.25(a), 1.4.25(b), 1.4.26(a), and 1.4.26(b) show that $\boldsymbol{X}_2$ and $\boldsymbol{Y}$ rapidly reach GS, although the initial condition (1.4.44) has a perturbation.

The evolution of state variables $t - x_{1\,11,11}, t - x_{2\,11,11}, t - x_{3\,11,11}, t - y_{1\,11,11}, t - y_{2\,11,11}$, and $t - y_{3\,11,11}$ are shown in Fig. 1.4.27. The evolution of chaotic patterns of the state variables $x_{k\,i,j}$, $y_{k\,i,j}$ over the time interval [0, 5] are shown in Fig. 1.4.28. It can be observed that the randomly perturbed initial patterns ($t = 0$) are evolving irregularly, but the GS relationships between $x_{l\,i,j}$ and $y_{l\,i,j}, l = 1, 2$ cannot been seen.

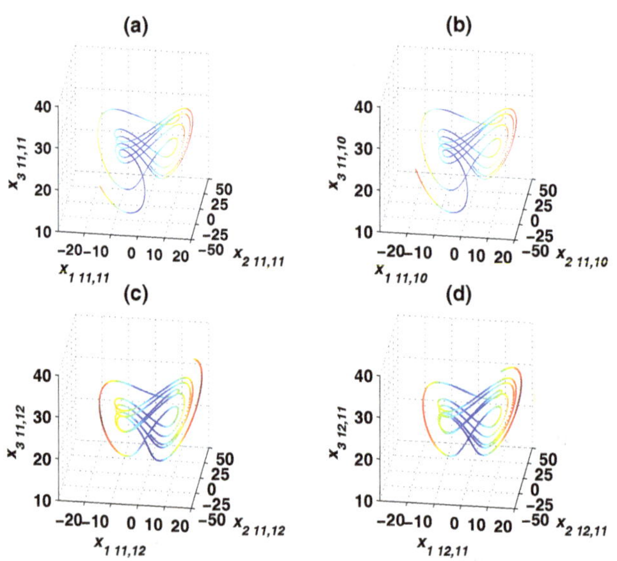

Fig. 1.4.24: Chaotic trajectories of some components of the state variables: (a) $x_{1\,12,11} - x_{2\,12,11} - x_{3\,12,11}$, (b) $x_{1\,12,10} - x_{2\,12,10} - x_{3\,12,10}$, (c) $x_{1\,12,12} - x_{2\,12,12}$ - $x_{3\,12,12}$ and (d) $x_{1\,13,11} - x_{2\,13,11} - x_{3\,13,11}$.

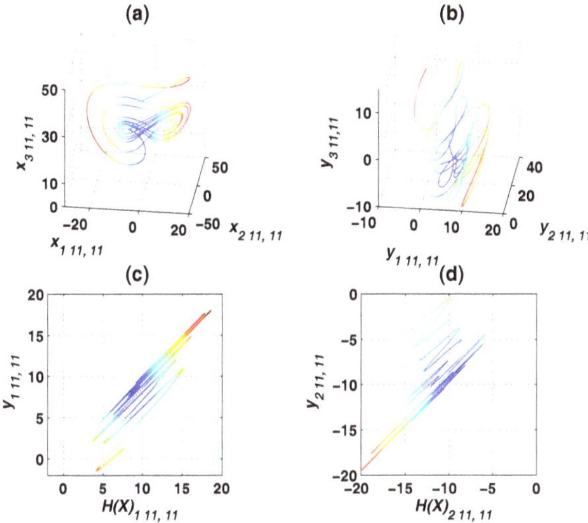

Fig. 1.4.25: GS trajectories of the components of the state variables: (a) $x_{1\,11,11} - x_{2\,11,11} - x_{3\,11,11}$, (b) $y_{1\,11,11} - y_{2\,11,11} - x_{3\,1,1}$. (c) $H(\mathbf{X})_{1\,11,11}(k)$ and $y_{1\,11,11}(k)$ are in GS, (d) $H(\mathbf{X})_{2\,11,11}(k)$ and $y_{2\,11,11}(k)$ are in GS.

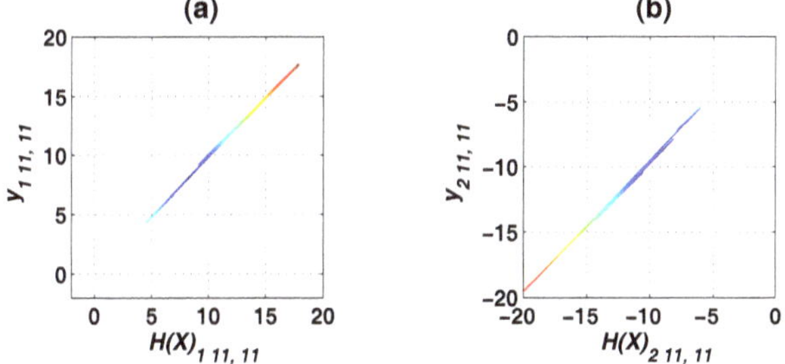

Fig. 1.4.26: After time is larger than 3, (a) and (b) show that the variables $x_{1\,11,11}, x_{2\,11,11}$ and $y_{1\,11,11}, y_{2\,11,11}$ are in GS with respect to the transformation H defined by (1.4.40)–(1.4.41).

The three-dimensional views of the evolution of the Chen CNN at different times are shown in Fig. 1.4.29, in which chaotic waves can been seen clearly. It can be observed that the irregular chaotic waves shown in the first three columns in Fig. 1.4.29 have been transformed to wall-shaped chaotic wave forms shown in the last three columns in Fig. 1.4.29.

In summary, the autonomous bidirectional Chen CNN has extremely complex dynamic behaviors.

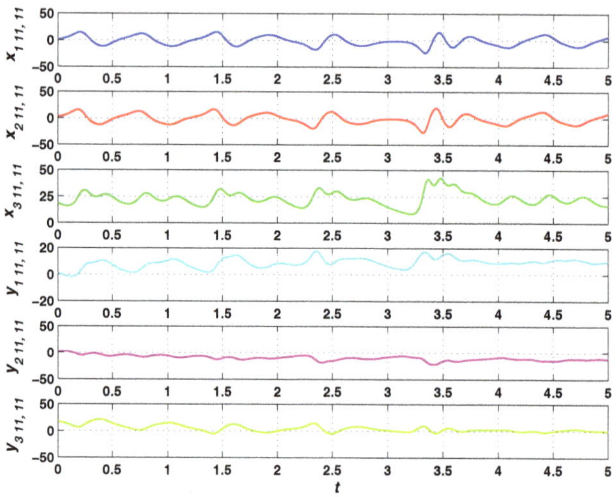

Fig. 1.4.27: The evolution of state variables: $t - x_{1\,11,11}, t - x_{2\,11,11}$, $t - x_{3\,11,11}, t - y_{1\,11,11}, t - y_{2\,11,11}$, and $t - y_{3\,11,11}$.

1.4.6 *Application of PGS Theorem to Non-autonomous Bidirectional CDADS*

In this subsection, an application of Theorem 1.10 to non-autonomous bidirectional CDADS is discussed.

First, a continuously differentiable and invertible transformation H is constructed, as follows:

$$H = T \circ \boldsymbol{B} \circ \tilde{H} : \mathbb{R}^{2\times M\times M} \times \mathbb{Z}^{+} \to \mathbb{R}^{2\times M\times M}, \tag{1.4.45}$$

where

$$\tilde{H} = (\tilde{h}_1, \tilde{h}_2) : \mathbb{R}^{2\times M\times M} \to \mathbb{R}^{2\times M\times M}, \tag{1.4.46}$$

$$\boldsymbol{B} : \mathbb{R}^2 \to \mathbb{R}^2, \tag{1.4.47}$$

and

$$T : \mathbb{R}^{2\times M\times M} \times \mathbb{Z}^{+} \to \mathbb{R}^{2\times M\times M}, \tag{1.4.48}$$

such that

$$\begin{aligned}\tilde{h}_1((x_{1\,i,j})_{21\times 21}) &= (\alpha^1_{i,j})_{21\times 21}(x_{1\,i,j})_{21\times 21}\\ &= (\tilde{x}_{1\,i,j})_{21\times 21},\end{aligned} \tag{1.4.49}$$

$$\begin{aligned}\tilde{h}_2((x_{2\,i,j})_{21\times 21}) &= (\alpha^2_{i,j})_{21\times 21}(x_{2\,i,j})_{21\times 21}\\ &= (\tilde{x}_{2\,i,j})_{21\times 21},\end{aligned} \tag{1.4.50}$$

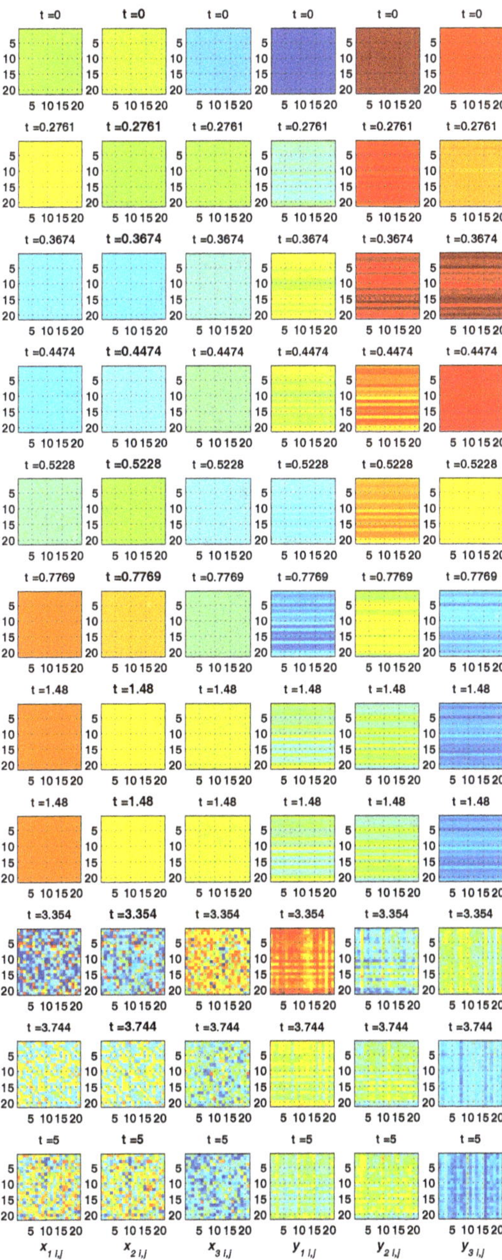

Fig. 1.4.28: Evolution of the patterns of the non-autonomous bidirectional Chen CNN over the time interval [0, 5].

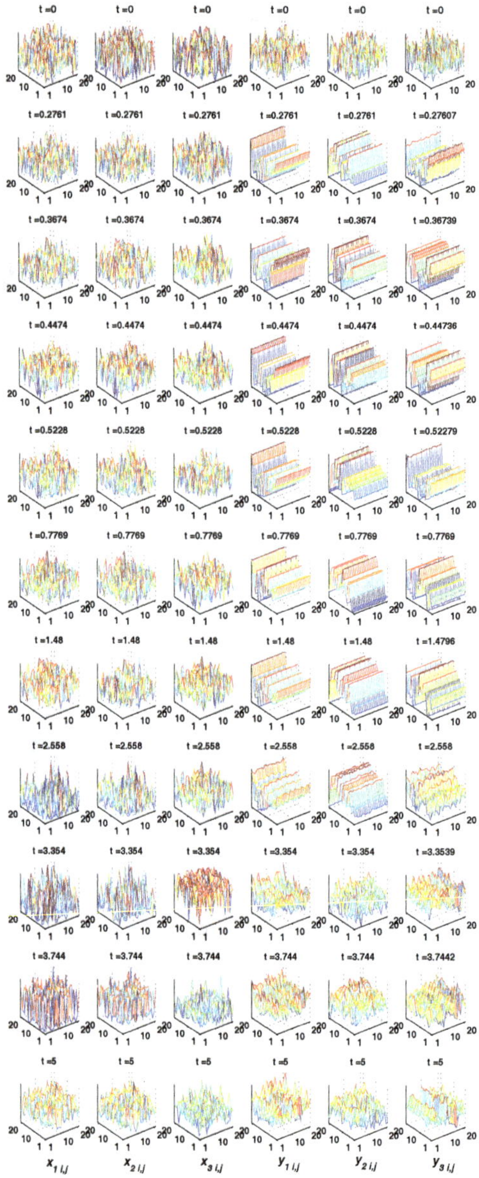

Fig. 1.4.29: The three-dimensional views of the non-autonomous bidirectional Chen CNN spiral waves at different time t. The vertical axes represent the state variables $x_{1\,i,j}$, $x_{2\,i,j}$, $x_{3\,i,j}$, $y_{1\,i,j}$, $y_{2\,i,j}$,$y_{3\,i,j}$, while the horizontal axes are the plane coordinates (i, j).

and, for any pair $(\tilde{x}_{1\,i,j}, \tilde{x}_{2\,i,j})$,

$$\boldsymbol{B}(\tilde{x}_{1\,i,j}, \tilde{x}_{2\,i,j}) = \begin{bmatrix} \dfrac{\tilde{x}_{1\,i,j}}{\tilde{x}_{1\,i,j}^2 + \tilde{x}_{2\,i,j}^2} & \dfrac{\tilde{x}_{2\,i,j}}{\tilde{x}_{1\,i,j}^2 + \tilde{x}_{2\,i,j}^2} \end{bmatrix}^{\mathrm{T}}, \tag{1.4.51}$$

and, for any pair $(\mathbf{X}, k) \in \mathbb{R}^{2\times M\times M} \times \mathbb{Z}^+$,

$$T(\mathbf{X}, k) = \mathbf{X} + 0.01\arctan(k). \tag{1.4.52}$$

This transformation, H is similar to the one shown in Fig. 1.4.13, but it has a term depending on the iterative steps k.

Using Theorem 1.10, the transformation defined by (1.4.45)–(1.4.52), and the modified 3D Lorenz chaotic map [Sprot (2003)], one can construct a PGS of non-autonomous bidirectional Lorenz CNN.

The first part of this Lorenz CNN has the following form:

$$\begin{cases} x_{1\,i,j}(k+1) = x_{1\,i,j}(k)x_{2\,i,j}(k) - x_{3\,i,j}(k) \\ \qquad\qquad +k_0\sin(\pi k/4) \\ \qquad\qquad +k_0\sin(\pi y_{1\,i,j}(k)y_{2\,i,j}(k)y_{3\,i,j}(k)) \\ x_{2\,i,j}(k+1) = x_{1\,i,j}(k) \\ x_{3\,i,j}(k+1) = x_{2\,i,j}(k) + D_2[x_{3i+1,j} + x_{3i-1,j} \\ \qquad\qquad +x_{3i,j+1} + x_{3i,j-1} - 4x_{4i,j}], \\ \qquad\qquad i,j = 1,2,\cdots 21, \end{cases} \tag{1.4.53}$$

where

$$k_0 = 10^{-6}, D = 10^{-8}. \tag{1.4.54}$$

Written in a compact form, it is

$$\mathbf{X}(k+1) = F(\ \mathbf{X}(k),\ \ \mathbf{Y}(k), k). \tag{1.4.55}$$

The second part of the above CCADS has the following form:

$$\mathbf{Y}(k+1) = \begin{bmatrix} H(F_2(\mathbf{X}(k), \mathbf{Y}(k), k), k+1) - Q(\mathbf{Y}, \mathbf{X}, k) \\ -(x_{4\,i,j})_{21\times 21} \end{bmatrix} \tag{1.4.56}$$

where

$$Q(\mathbf{Y}, \mathbf{X}) = \frac{1}{10}\mathbf{e}.$$

Now, select the following initial conditions:

$$(x_{l\,i,j}(0))_{21\times 21} = \mathbf{X}(l) + 0.02(rand(21,21) - 0.5), \quad l = 1,2,3,$$

$$\mathbf{Y}_2(0) = H(\mathbf{X}_2(0), 0) + 0.2(rand(2,21,21) - 0.5),$$

where $rand(2, 21, 21)$ is a Matlab command.

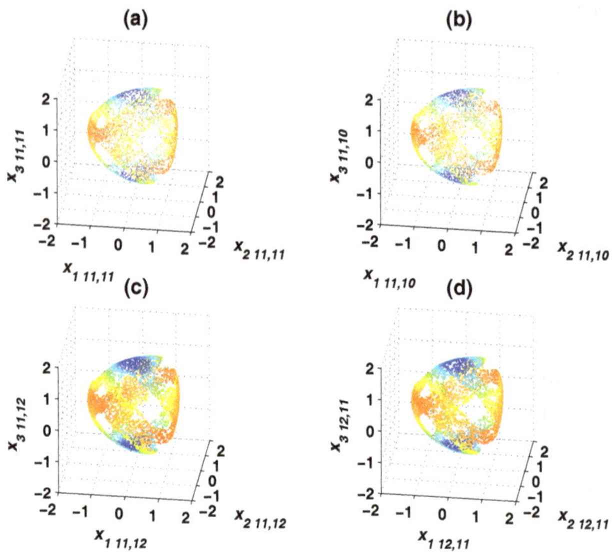

Fig. 1.4.30: Chaotic trajectories of some components of the state variables: (a) $x_{1\,11,11} - x_{2\,11,11} - x_{3\,11,11}$, (b) $x_{1\,11,10} - x_{2\,11,10} - x_{3\,11,10}$, (c) $x_{1\,11,12} - x_{2\,11,12} - x_{3\,11,12}$ and (d) $x_{1\,12,11} - x_{2\,12,11} - x_{3\,12,11}$.

The chaotic orbits of some components $x'_{l\,i,j}$ of the state variables $\boldsymbol{X}$ of the first 5,000 iterations are shown in Figs. 1.4.30(a)–(d). It can be observed that the dynamic behaviors of the neighboring cells at the lattice: (11, 11), (11, 10), (11, 12) and (12, 11), are completely different.

The chaotic trajectories of the components $x_{k\,11,11}$ and $y_{k\,11,11}$ of the state variables $\boldsymbol{X}$ and $\boldsymbol{Y}$ over the first 5,000 iterations are shown in Fig. 1.4.31(a) and (b). Figures 1.4.31(c) and (d) show that, although the initial condition (1.4.57) has a perturbation, $\boldsymbol{X}_2$ and $\boldsymbol{Y}_2$ rapidly reach GS, as the theory predicts.

The evolution of state variables: $t - x_{1\,11,11}$, $t - x_{2\,11,11}$, $t - x_{3\,11,11}$, $t - y_{1\,11,11}$, $t - y_{2\,11,11}$, and $t - y_{3\,11,11}$ are shown in Fig. 1.4.32. However, the GS relationships between $x_{l\,i,j}$ and $y_{l\,i,j}, l = 1, 2$ cannot be seen.

The evolution of the chaotic patterns of the state variables $x_{k\,i,j}$, $y_{k\,i,j}$ over the first 5,000 iterations are shown in Fig. 1.4.33. It can be observed that the randomly perturbed initial patterns are evolving irregularly, but the GS relationships between $x_{l\,i,j}$ and $y_{l\,i,j}, l = 1, 2$, cannot be seen.

The three-dimensional views of the evolution of the Lorenz CNN at different iterative steps k are shown in Fig. 1.4.34, in which chaotic waves

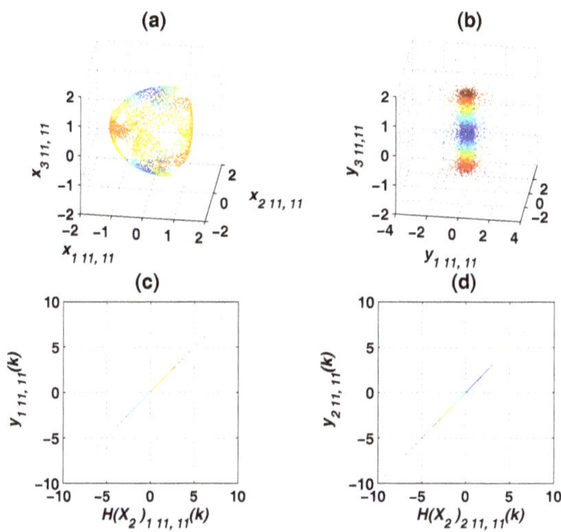

Fig. 1.4.31: Chaotic GS trajectories of the components of the state variables: (a) $x_{1\,11,11} - x_{2\,11,11} - x_{3\,11,11}$, (b) $y_{1\,11,11} - y_{2\,11,11} - y_{3\,1,1}$, (c) $H(\mathbf{X}_2)_{1\,11,11}(k)$ and $y_{1\,11,11}(k)$ are in GS, (d) $H(\mathbf{X}_2)_{2\,11,11}(k)$ and $y_{2\,11,11}(k)$ are in GS.

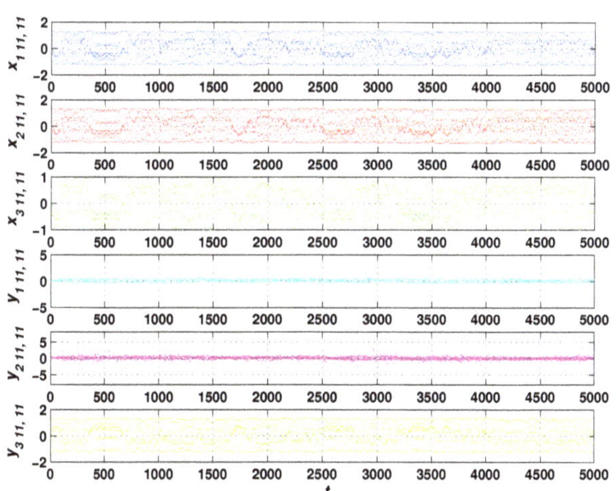

Fig. 1.4.32: The evolution of state variables: $t - x_{1\,11,11}$, $t - x_{2\,11,11}$, $t-x_{3\,11,11}$, $t - y_{1\,11,11}$, $t - y_{2\,11,11}$, and $t - y_{3\,11,11}$

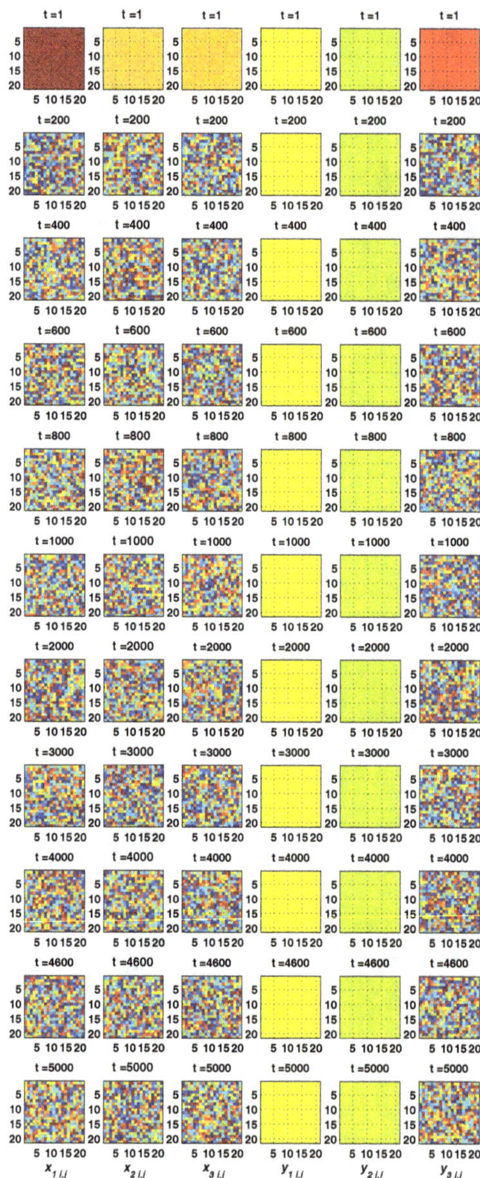

Fig. 1.4.33: Evolution of the patterns of the discrete non-autonomous bidirectional Lorenz CNN over the first 5,000 iterations.

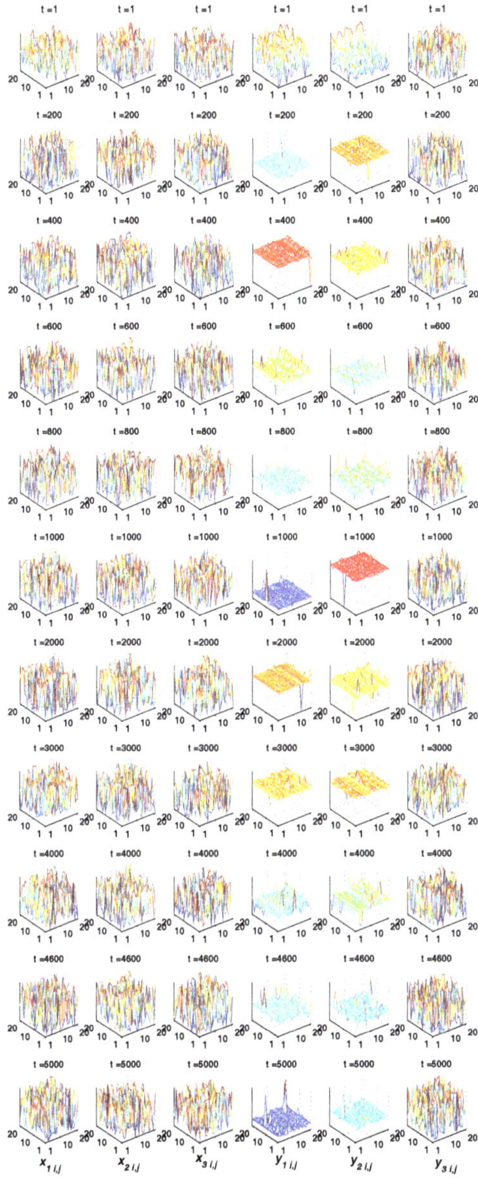

Fig. 1.4.34: The three-dimensional views of the discrete non-autonomous bidirectional Lorenz CNN spiral waves at different iteration steps k. The vertical axes represent the state variables $x_{1\,i,j}$, $x_{2\,i,j}$, $x_{3\,i,j}$, $y_{1\,i,j}$, $y_{2\,i,j}$, $y_{3\,i,j}$, while the horizontal axes are the plane coordinates (i, j).

are seen clearly. It can be observed that the irregular chaotic waves shown in the 1st-2nd columns in Fig. 1.4.34 have been transformed to plane-shaped chaotic wave forms shown in the 4th–5th columns in Fig. 1.4.34.

In summary, the non-autonomous bidirectional Lorenz CNN has extremely complex dynamic behaviors.

1.4.7 *Application of PGS Theorem to Non-autonomous Bidirectional CCADS*

In this subsection, an application of Theorem 1.11 to non-autonomous bidirectional CCADS is discussed.

Based on the hyperchaotic Chen system proposed by Jia *et al.* [Jia *et al.* (2010)], a non-autonomous bidirectional Chen CNN is constructed with eight state variables in $8 \times 21 \times 21$ dimensions.

First, a continuously differentiable transformation H is constructed, as follows:

$$H = T \circ \boldsymbol{B} \circ \tilde{H} : \mathbb{R}^{3\times 21\times 21} \times \mathbb{Z}^{+} \to \mathbb{R}^{3\times 21\times 21}, \tag{1.4.57}$$

where

$$\tilde{H} = (\tilde{h}_1, \tilde{h}_2, \tilde{h}_3) : \mathbb{R}^{3\times 21\times 21} \to \mathbb{R}^{3\times 21\times 21}, \tag{1.4.58}$$

$$\boldsymbol{B} : \mathbb{R}^3 \to \mathbb{R}^3, \tag{1.4.59}$$

and

$$T : \mathbb{R}^{3\times 21\times 21} \times \mathbb{R}^{+} \to \mathbb{R}^{3\times 21\times 21}, \tag{1.4.60}$$

such that

$$\begin{aligned}
\tilde{h}_1((x_{1\,i,j})_{21\times 21}) &= (\alpha^1_{i,j})_{21\times 21}(x_{1\,i,j})_{21\times 21} \\
&= (\tilde{x}_{1\,i,j})_{21\times 21}, &(1.4.61)\\
\tilde{h}_2((x_{2\,i,j})_{21\times 21}) &= (\alpha^2_{i,j})_{21\times 21}(x_{2\,i,j})_{21\times 21} \\
&= (\tilde{x}_{2\,i,j})_{21\times 21}, &(1.4.62)\\
\tilde{h}_3((x_{3\,i,j})_{21\times 21}) &= (\alpha^3_{i,j})_{21\times 21}(x_{3\,i,j})_{21\times 21} \\
&= (\tilde{x}_{3\,i,j})_{21\times 21}, &(1.4.63)
\end{aligned}$$

where

$$\begin{cases} \mathbf{A}_1 = (\alpha^1_{i,j})_{21\times 21} \\ \mathbf{A}_2 = (\alpha^2_{i,j})_{21\times 21} \\ \mathbf{A}_3 = (\alpha^3_{i,j})_{21\times 21} \end{cases} \tag{1.4.64}$$

are invertible matrices, and, for any triple $(\tilde{x}_{1\,i,j}, \tilde{x}_{2\,i,j}, \tilde{x}_{2\,i,j})$,

$$\boldsymbol{B}(\tilde{x}_{1\,i,j}, \tilde{x}_{2\,i,j}, \tilde{x}_{3\,i,j}) = \begin{bmatrix} -1 & 1 & 1 \\ 1 & 1 & 1 \\ 1 & -1 & 1 \end{bmatrix} \begin{bmatrix} \arctan(\tilde{x}_{1\,i,j}) \\ \arctan(\tilde{x}_{2\,i,j}) \\ \arctan(\tilde{x}_{3\,i,j}) \end{bmatrix}, \quad (1.4.65)$$

and, for any pair $(\boldsymbol{X}_3, k) \in \mathbb{R}^{3\times21\times21} \times \mathbb{R}^+$,

$$T(\boldsymbol{X}_3, k) = \boldsymbol{X}_3 + 10^{-4}\tanh(t). \quad (1.4.66)$$

This transformation H is similar to the one shown in Fig. 1.4.13, but it has a term depending on time t.

Now, using Theorem 1.11, the transformation defined by (1.4.57)–(1.4.60), and the modified Chen system [Jia *et al.* (2010)], one can construct a PGS non-autonomous bidirectional CCADS.

The first part of the Chen CNN has the following form:

$$\begin{cases} \dot{x}_{1i,j} = a(x_{2i,j} - x_{1i,j}) + k_0\cos(\pi t) \\ \dot{x}_{2i,j} = 4x_{1i,j} + cx_{2i,j} \\ \qquad -10x_{1i,j}x_{3i,j} + 4x_{4i,j} \\ \dot{x}_{3i,j} = x_{2i,j}^2 - bx_{3i,j} \\ \qquad +k_0\arctan(y_{1i,j}y_{2i,j}y_{3i,j}y_{4i,j}) \\ \dot{x}_{4i,j} = -dx_{1i,j} + D[x_{4i+1,j} + x_{4i-1,j} \\ \qquad +x_{4i,j+1} + x_{4i,j-1} - 4x_{4i,j}], \\ \qquad i,j = 1,2,\ldots,21, \end{cases} \quad (1.4.67)$$

where $a = 35, b = 3, c = 21$, $d = 2, k_0 = 10^{-4}$, $D = 5\times10^{-4}$.

In a compact form, the CCADS (1.4.67) can be written as

$$\dot{\mathbf{X}} = F(\mathbf{X}, \mathbf{Y}, t). \quad (1.4.68)$$

The second part of the CCDAS has the following form:

$$\dot{\mathbf{Y}} = G(\mathbf{Y}, \mathbf{X}, t), \quad (1.4.69)$$

where

$$
\begin{aligned}
G_{l\,i,j}(\boldsymbol{Y},\boldsymbol{X},t) &= \sum_{l'=1}^{3}\sum_{i'=1}^{21}\sum_{j'=1}^{21}\frac{\partial h_{l\,i,j}(\boldsymbol{X}_3,t)}{\partial x_{l'\,i',j'}} \\
&\quad \times f_{l'\,i',j'}(\mathbf{X},\,\mathbf{Y},t)+\frac{\partial h_{l\,i,j}(\mathbf{X}_3 t)}{\partial t} \\
&\quad -q_{l\,i,j}(\boldsymbol{Y},\boldsymbol{X},t), \\
&= \frac{1}{1+\left(\sum_{k=1}^{21}\alpha_{i,k}^{l}x_{lk,j}\right)^2}\sum_{k=1}^{21}\alpha_{i,k}^{l} \\
&\quad \times f_{lk,j}(\mathbf{X},\,Y,t)+10^{-4}(1-\tanh(t)^2) \\
&\quad -q_{l\,i,j}(\boldsymbol{Y},\boldsymbol{X},t), \qquad (1.4.70) \\
&\quad l=1,2,3,i,j=1,2,\cdots,21, \\
G_{4\,i,j}(\boldsymbol{Y},\boldsymbol{X},t) &= x_{4\,i,j}, \qquad (1.4.71)
\end{aligned}
$$

$$
q_{l\,i,j}(\boldsymbol{Y},\boldsymbol{X},t) = -10(h_{l\,i,j}(\boldsymbol{X}_3)-y_{l\,i,j}). \qquad (1.4.72)
$$

From Theorem 1.11, systems (1.4.68) and (1.4.69) are in GS with respect to the transformation H defined by (1.4.57)–(1.4.66) (see Fig. 1.4.13).

Now, select the following initial conditions:

$$
\begin{aligned}
(x_{l\,i,j}(0))_{21\times 21} &= \mathbf{X}_0(l)+0.02(rand(21,21)-0.5), l=1,2,3, \\
\boldsymbol{Y}(0) &= \boldsymbol{X}(0)+0.02(rand(21,21)-0.5), \qquad (1.4.73)
\end{aligned}
$$

where $\mathbf{X}_0 = [-3.6462,-4.726,3.0802,3.0802]$, and $rand(21,21)$ is a Matlab command.

The chaotic orbits of some components $x_{l\,i,j}$ of the state variables $\boldsymbol{X}$ over the time interval [0, 8] are shown in Figs. 1.4.35(a)–(d).

The chaotic orbits of some components $y_{l\,i,j}$ of the state variables $\boldsymbol{Y}$ over the time interval [0, 8] are shown in Figs. 1.4.36(a)–(d). It can be observed that the transformation H compresses the butterfly-shaped trajectories (Fig. 1.4.35) to prism-shaped trajectories (Fig. 1.4.36).

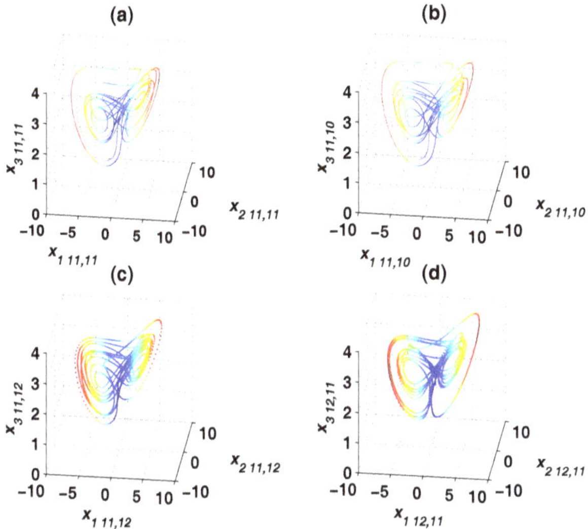

Fig. 1.4.35: Chaotic trajectories of some components of the state variables $\mathbf{X}(t)$: (a) $x_{1\,11,11} - x_{2\,11,11} - x_{3\,11,11}$, (b) $x_{1\,11,10} - x_{2\,11,10} - x_{3\,11,10}$, (c) $x_{1\,11,12} - x_{2\,11,12} - x_{3\,11,12}$, (d) $x_{1\,12,11} - x_{2\,12,11} - x_{3\,12,11}$.

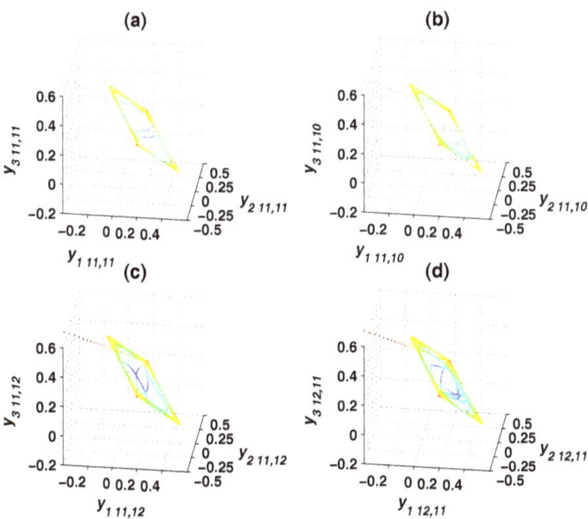

Fig. 1.4.36: Chaotic trajectories of some components of the state variables $\boldsymbol{Y}(t)$: (a) $y_{1\,11,11} - y_{2\,11,11} - y_{3\,11,11}$, (b) $y_{1\,11,10} - y_{2\,11,10} - y_{3\,11,10}$, (c) $y_{1\,11,12} - y_{2\,11,12} - y_{3\,11,12}$, (d) $y_{1\,12,11} - y_{2\,12,11} - y_{3\,12,11}$.

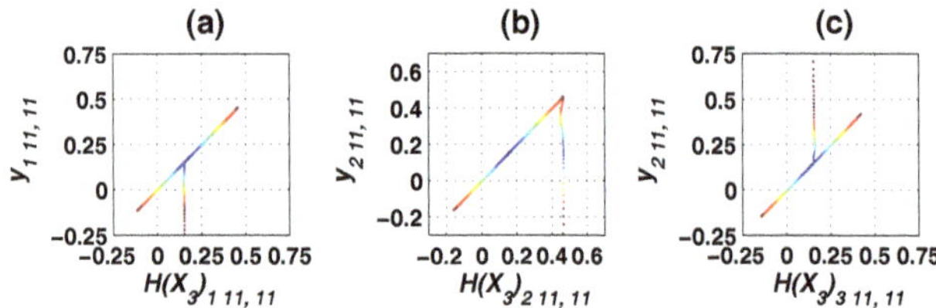

Fig. 1.4.37: The state variables $\mathbf{X}_3$ and $\mathbf{Y}_3$ are in GS with respect to the transformation H. (a) $H(\mathbf{X}_3)_{1\,11,11}$-$y_{1\,11,11}$, (b) $H(\mathbf{X}_3)_{2\,11,11}$- $y_{2\,11,11}$, (c) $H(\mathbf{X}_3)_{3\,11,11}$-$y_{3\,11,11}$.

Figure 1.4.37 shows that, although there are initial perturbations (1.4.73), the state variables $\boldsymbol{X}$ and $\boldsymbol{Y}$ reach GS rapidly.

The evolution of state variables: $t - x_{1\,11,11}$, $t - x_{2\,11,11}$, $t - x_{3\,11,11}$, and $t - x_{4\,11,11}$ over the time interval [0, 8] are shown in Fig. 1.4.38. The evolution of state variables: $t-y_{1\,11,11}$, $t-y_{2\,11,11}$, $t-y_{3\,11,11}$, and $t-y_{4\,11,11}$ are shown in Fig. 1.4.39. However, the GS relationships between $x_{l\,i,j}$ and $y_{l\,i,j}, l = 1,2,3$ cannot be seen.

The evolution of the chaotic patterns of the state variables $x_{k\,i,j}$, $y_{k\,i,j}$ over the time interval [0, 8] are shown in Fig. 1.4.40. It can be observed that the small randomly perturbed initial patterns ($t = 0$) are evolving irregularly, but the GS relationships between $x_{l\,i,j}$ and $y_{l\,i,j}, l = 1,2,3$, cannot be seen.

The three-dimensional views of the evolution of the Chen CNN at different times are shown in Fig. 1.4.41, in which chaotic waves can be seen clearly. It can be observed that the irregular chaotic waves shown in the first three columns in Fig. 1.4.41 have been transformed to the wall-shaped chaotic wave forms shown in the 5th-7th columns in Fig. 1.4.41.

In summary, the non-autonomous bidirectional Chen CNN has extremely complex dynamic behaviors.

1.5 Conclusions

The main results presented in this part include seven constructive GS theorems, which describe the general forms of two autonomous or non-autonomous CDADS and CCADS to be in GS with respect to a transformations H. These theorems (Theorems 1.3, 1.4, 1.7–1.11) have extended some existing results on GS of vector differential systems [Zhang and Min (2000)], vector discrete difference systems [Zang *et al.* (2007)], array of

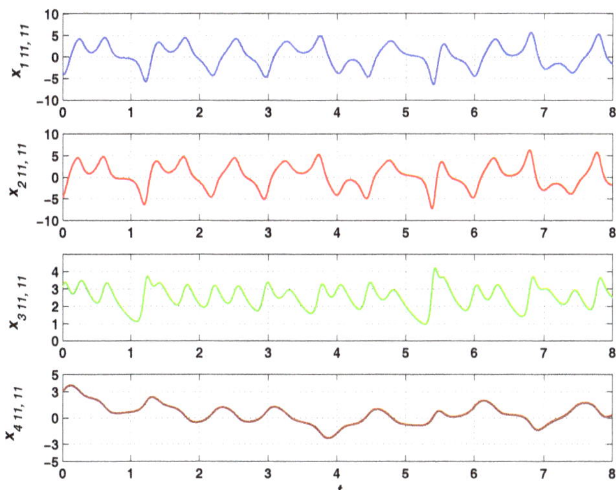

Fig. 1.4.38: The evolution of state variables: $t - x_{1\,11,11}$, $t - x_{2\,11,11}$, $t - x_{3\,11,11}$, and $t - x_{4\,11,11}$.

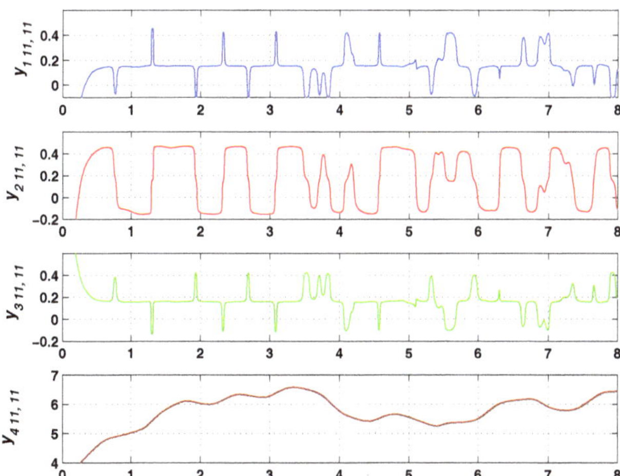

Fig. 1.4.39: The evolution of state variables: $t - y_{1\,11,11}$, $t - y_{2\,11,11}$, $t - y_{3\,11,11}$, and $t - y_{4\,11,11}$.

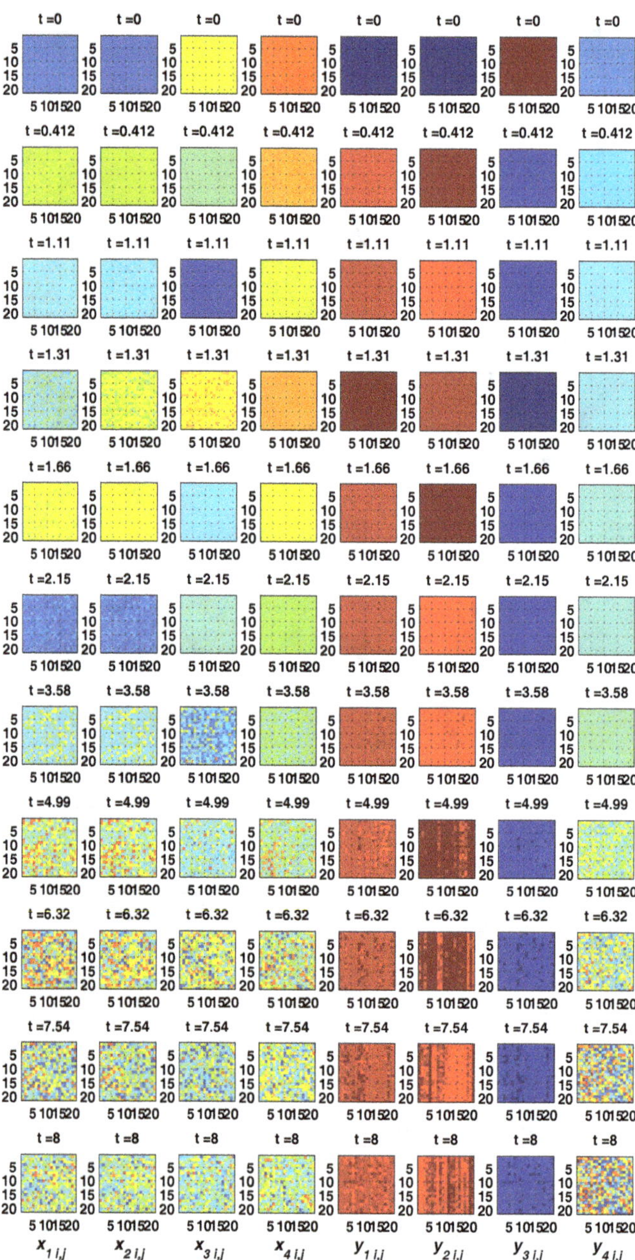

Fig. 1.4.40: Evolution of the patterns of the non-autonomous bidirectional Chen CNN over the time interval [0, 8].

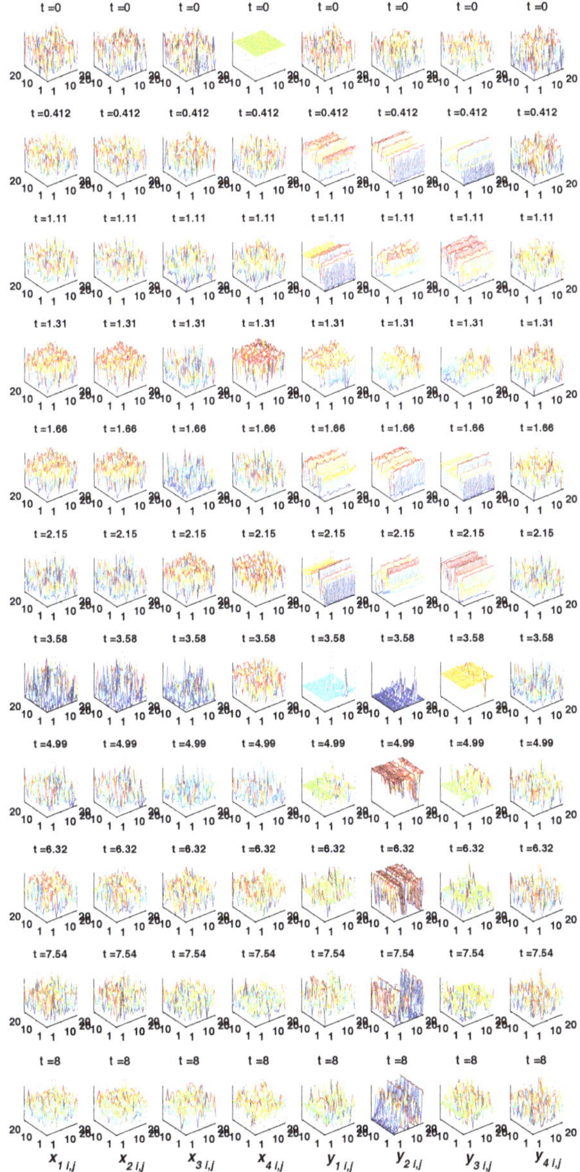

Fig. 1.4.41: The three-dimensional views of the non-autonomous bidirectional Chen CNN spiral waves at different time t. The vertical axes represent the state variables $x_{1\,i,j}$, $x_{2\,i,j}$, $x_{3\,i,j}$, $x_{4\,i,j}$, $y_{1\,i,j}, y_{2\,i,j}$, $y_{3\,i,j}$, $y_{4\,i,j}$, while the horizontal axes are the plane coordinates (i,j).

discrete difference systems [Zang and Min (2008)], bidirectional vector continuous differential systems and discrete difference systems [Ji *et al.* (2008)], array of differential systems [Min and Zang (2009)], bidirectional array of differential as well as difference systems [Zang *et al.* (2012)].

The new theorems confirm some equivalent classes of GS systems. The difference between two GS systems in each equivalent class is a function that makes the zero solution of the error system be asymptotically stable.

The new approach developed in this part is to use the GS theorems to construct GS CDADS and CCADS, rather than to use the classic Lyapunov exponent techniques in the literature. Generally speaking, Lyapunov exponents of CDADS and CCADS are very difficult to calculate, therefore the new alternative developed here should be welcome.

The other new results include seven new GS CNN. Numerical simulations have shown the complex dynamical behaviors of such GS CNN.

Based on the eleven new GS theorems on CDADS and CCADS and the seven examples, one can better understand why many different systems can be in various forms of GS. The GS theorems presented in this monograph should be useful for analyzing GS phenomena of many practical systems in different fields.

Chapter 2

Generalized Consensus in an Array of Nonlinear Dynamic Systems with Applications to Chaotic CNN

2.1 Introduction

This part of the monograph investigates the generalized consensus (GC, or called generalized stability [Min and Chen (2017)]) of coupled arrays of dynamical systems, in a way parallel to the study of the generalized synchronization carried out in the first part of the book above.

Conceptually, synchronization means that the dynamical behaviors of two system arrays trend to be identical as time evolves, while consensus means that the difference between the dynamics of the two arrays becomes sufficiently small when the processing time is long enough, where the latter becomes the former if the small difference actually becomes zero.

The study of consensus is application-driven, where coordinative multi-agent systems are expected to move to be close (so as to perform cooperative tasks) but not exactly getting together (so as to avoid crashing). There are many practical scenarios where consensus is required (for example see [Subbiah *et al.* (1993); Hirche and Hara (2008); Takaba (2011); Antal *et al.* (2014); Belotti *et al.* (2019); Panitz and Glückler (2020); de Oliveira *et al.* (2020)]), and there are many works devoted to the investigation of this subject (mainly on systems or networks of systems (for example see [Zhang and Zhuang (2019); Semin and Levin (2019); Samli *et al.* (2019); Panitz and Glückler (2020); Brewer *et al.* (2020); Chlebus *et al.* (2020)]), but notably not on coupled arrays of systems).

Chaos generalized synchronization (GS) is one of the focal research topics in nonlinear science. The concepts of GC and GS are the extensions of the concept of synchronization. A general question is: if two systems can achieve GC with respect to a transformation, what kind of representations should these systems have? To answer this question, in previous researches we derived some constructive constructive GC theorems on discrete and differentiable vector systems [Wang *et al.* (2015a,b); Zhang *et al.* (2015a,b); Yang *et al.* (2015)].

The rest part of this Chapter is devoted to a detailed study of generalized consensus in an array of nonlinear dynamic systems, for both CDADS and CCADS, with applications to chaotic CNN. Section 2.2 introduces some definitions on space matrices, the GC of CDADS, the GC of CCADS, and their associated transformations. Section 2.3 presents GC theorems for coupled arrays of difference and differentiable systems, respectively. As application examples, six autonomous or non-autonomous GC CDADS and GC CCADS are simulated and analyzed in Section 2.4. Finally, some concluding remarks are given in Section 2.5.

2.2 Basic Concepts and Definitions

To establish new generalized consensus (GC) theorems for CDADS and CCADS, some concepts and definitions are first introduced [Min and Zang (2009); Min and Chen (2013)].

For each $l = 1, 2, ..., n$, let

$$\mathbf{X}^l = (x_{l\,i,j})_{M\times N}$$

be an $M \times N$ matrix. Then, for convenience,

$$\begin{aligned}\mathbf{X} &= (\mathbf{X}^1, \mathbf{X}^2, \cdots, \mathbf{X}^n)^{\mathrm{T}} \\ &\triangleq (x_{l\,i,j})_{n\times M\times N}\end{aligned}$$

is called a space matrix, with elements $x_{l\,i,j}$, $i = 1, ..., M$, $j = 1, ..., N$, and $l = 1, 2, ..., n$.

The space of all space matrices is the set of all real space matrices $(x_{l\,i,j})_{n\times M\times N}$, denoted as $\mathbb{R}^{n\times M\times N}$. Define

$$\begin{aligned}\mathbf{X} + \mathbf{Y} &= (x_{l\,i,j} + y_{l\,i,j})_{n\times M\times N}, \\ \alpha\mathbf{X} &= (\alpha x_{l\,i,j})_{n\times M\times N}, \quad \alpha \in \mathbb{R}.\end{aligned}$$

Then, $\mathbb{R}^{n\times M\times N}$ is a real vector space of dimension $n \times M \times N$. Since all norms on a finite-dimensional space are equivalent, one can define any norm $\|\cdot\|$ on it.

Let $H : \mathbb{R}^{n\times M\times N} \to \mathbb{R}^{n\times M\times N}$ be a transformation (see Fig. 1.2.1). Then, for every $\mathbf{X} \in \mathbb{R}^{n\times M\times N}$, one can write

$$\begin{aligned}H(\mathbf{X}) = (h_{l\,i,j}(\mathbf{X})), l &= 1, \cdots, n, \\ i = 1, \cdots, M,\ j &= 1, \cdots, N,\end{aligned}$$

or

$$H(\mathbf{X}) = (h_{l\,i,j}(\mathbf{X}))_{n\times M\times N}.$$

2.2.1 *GC of an Array of Difference Systems*

Definition 2.1. The CDADS defined by (1.2.5) and (1.2.6) are said to be in Generalized Consensus with respect to a Transformation $H : \mathbb{R}^{m\times M\times N} \to \mathbb{R}^{m\times M\times N}$ (see Fig. 1.2.2), abbreviated as GC, if for all $\epsilon > 0$, there exists an open subset $B_\epsilon \subset \mathbb{R}^{n\times M\times N} \times \mathbb{R}^{m\times M\times N}$, such that for any trajectory $(\boldsymbol{X}(k), \boldsymbol{Y}(k))$ of systems (1.2.5) and (1.2.6) with initial condition $(\boldsymbol{X}(0), \boldsymbol{Y}(0)) \in B_\epsilon$, the following property holds:

$$\|H(\mathbf{X}_m(k)) - \mathbf{Y}(k)\| < \epsilon \quad k = 1, 2, \cdots \tag{2.2.1}$$

where

$$\begin{aligned}\mathbf{X}_m(k) &= (\mathbf{X}^1(k), \cdots, \mathbf{X}^m(k))^{\mathrm{T}}\\ &= (x_{l\,i,j}(k))_{m\times M\times N},\\ \mathbf{Y}(k) &= (\mathbf{Y}^1(k), \cdots, \mathbf{Y}^m(k))^{\mathrm{T}}\\ &= (y_{l\,i,j}(k))_{m\times M\times N}.\end{aligned}$$

(2) GC Bidirectional CDADS

Definition 2.2. The bidirectional CDADS defined by (1.2.14) and (1.2.15) are said to be in Generalized Consensus with respect to a Transformation $H : \mathbb{R}^{m\times M\times N} \to \mathbb{R}^{m\times M\times N}$ (see Fig. 1.2.2). abbreviated as GC, if for all $\epsilon > 0$, there exists an open subset $B_\epsilon \subset \mathbb{R}^{n\times M\times N} \times \mathbb{R}^{m\times M\times N}$, such that for any trajectory $(\boldsymbol{X}(k), \boldsymbol{Y}(k))$ of systems (1.2.14) and (1.2.15) with initial condition $(\boldsymbol{X}(0), \boldsymbol{Y}(0)) \in B_\epsilon$, the following property holds:

$$\|H(\mathbf{X}_m(k)) - \mathbf{Y}(k)\| < \epsilon, k = 1, 2, \cdots \tag{2.2.2}$$

where

$$\begin{aligned}\mathbf{X}_m(k) &= (\mathbf{X}^1(k), \cdots, \mathbf{X}^m(k))^{\mathrm{T}}\\ &= (x_{l\,i,j}(k))_{m\times M\times N},\\ \mathbf{Y}(k) &= (\mathbf{Y}^1(k), \cdots, \mathbf{Y}^m(k))^{\mathrm{T}}\\ &= (y_{l\,i,j}(k))_{m\times M\times N}.\end{aligned}$$

(3) GC Non-autonomous CDADS

Definition 2.3. The non-autonomous CDADS defined by (1.2.23) and (1.2.24) are said to be in Generalized Consensus with respect to a Transformation $H : \mathbb{R}^{m\times M\times N} \times \mathbb{Z}^+ \to \mathbb{R}^{m\times M\times N}$ (see Fig. 1.2.3), abbreviated as

GC, if for all $\epsilon > 0$, there exists an open subset $B_\epsilon \subset \mathbb{R}^{n\times M\times N} \times \mathbb{R}^{m\times M\times N}$, such that for any trajectory $(\boldsymbol{X}(k), \boldsymbol{Y}(k))$ of systems (1.2.23) and (1.2.24) with initial condition $(\boldsymbol{X}(0), \boldsymbol{Y}(0)) \in B_\epsilon$, the following property holds:

$$\|H(\mathbf{X}_m(k), k) - \mathbf{Y}(k)\| < \epsilon, \ k = 1, 2, \cdots \tag{2.2.3}$$

where

$$\begin{aligned}\mathbf{X}_m(k) &= (\mathbf{X}^1(k), \cdots, \mathbf{X}^m(k))^{\mathrm{T}} \\ &= (x_{l\,i,j}(k))_{m\times M\times N}, \\ \mathbf{Y}(k) &= (\mathbf{Y}^1(k), \cdots, \mathbf{Y}^m(k))^{\mathrm{T}} \\ &= (y_{l\,i,j}(k))_{m\times M\times N}.\end{aligned}$$

Remark 2.2.1. As a special case, $H : \mathbb{R}^{m\times M\times N} \times \mathbb{Z}^+ \to \mathbb{R}^{m\times M\times N}$ may be independent of the iteration step $k \in \mathbb{Z}^+$.

(4) GC Non-autonomous bidirectional CDADS

Definition 2.4. The non-autonomous bidirectional CDADS defined by (1.2.32) and (1.2.33) are said to be in Generalized Consensus with respect to a Transformation $H : \mathbb{R}^{m\times M\times N} \times \mathbb{Z}^+ \to \mathbb{R}^{m\times M\times N}$ (see Fig. 1.2.3), if for all $\epsilon > 0$, there exists an open subset $B \subset \mathbb{R}^{n\times M\times N} \times \mathbb{R}^{m\times M\times N}$, such that for any trajectory $(\boldsymbol{X}(k), \boldsymbol{Y}(k))$ of systems (1.2.32) and (1.2.33) with initial condition $(\boldsymbol{X}(0), \boldsymbol{Y}(0)) \in B_\epsilon$, the following property holds:

$$\|H(\mathbf{X}_m(k), k) - \mathbf{Y}(k)\| < \epsilon, k = 1, 2, \cdots \tag{2.2.4}$$

where

$$\begin{aligned}\mathbf{X}_m(k) &= (\mathbf{X}^1(k), \cdots, \mathbf{X}^m(k))^{\mathrm{T}} \\ &= (x_{l\,i,j}(k))_{m\times M\times N}, \\ \mathbf{Y}(k) &= (\mathbf{Y}^1(k), \cdots, \mathbf{Y}^m(k))^{\mathrm{T}} \\ &= (y_{l\,i,j}(k))_{m\times M\times N}.\end{aligned}$$

Remark 2.2.2. As a special case, $H : \mathbb{R}^{m\times M\times N} \to \mathbb{R}^{m\times M\times N}$ may be independent of the iteration step $k \in \mathbb{Z}^+$.

2.2.2 *GC of an Array of Differential Systems*

(1) GC CCADS

Definition 2.5. The CCADS defined by (1.2.41) and (1.2.42) are said to be in Generalized Consensus with respect to a Transformation $H : \mathbb{R}^{m\times M\times N} \to \mathbb{R}^{m\times M\times N}$ (see Fig. 1.2.2), if for all $\epsilon > 0$, there exists an open subset $B_\epsilon \subset \mathbb{R}^{n\times M\times N}\times\mathbb{R}^{m\times M\times N}$, such that for any trajectory $(\mathbf{X}(t), \mathbf{Y}(t))$ of systems (1.2.41) and (1.2.42) with initial condition $(\mathbf{X}(0), \mathbf{Y}(0)) \in B_\epsilon$, the following property holds:

$$\|H(\mathbf{X}_m(t)) - \mathbf{Y}(t)\| < \epsilon, \forall t \geq 0 \tag{2.2.5}$$

where

$$\begin{aligned}\mathbf{X}_m(t) &= (x_{l\,i,j}(t))_{m\times M\times N},\\ \mathbf{Y}(t) &= (y_{l\,i,j}(t))_{m\times M\times N}.\end{aligned}$$

(2) GC Bidirectional CCADS

Definition 2.6. The bidirectional CCADS defined by (1.2.50) and (1.2.51) are said to be in Generalized Consensus respect to a Transformation $H : \mathbb{R}^{m\times M\times N} \to \mathbb{R}^{m\times M\times N}$ (see Fig. 1.2.2), abbreviated as GC, if for all $\epsilon > 0$, there exists an open subset $B_\epsilon \subset \mathbb{R}^{n\times M\times N} \times \mathbb{R}^{m\times M\times N}$, such that for any trajectory $(\mathbf{X}(t), \mathbf{Y}(t))$ of systems (1.2.50) and (1.2.51) with initial condition $(\mathbf{X}(0), \mathbf{Y}(0)) \in B_\epsilon$, the following property holds:

$$\|H(\mathbf{X}_m(t)) - \mathbf{Y}(t)\| < \epsilon, \ \forall t \geq 0 \tag{2.2.6}$$

where

$$\begin{aligned}\mathbf{X}_m(t) &= (x_{l\,i,j}(t))_{m\times M\times N},\\ \mathbf{Y}(t) &= (y_{l\,i,j}(t))_{m\times M\times N}.\end{aligned}$$

(3) GC Non-autonomous CCADS

Definition 2.7. A non-autonomous CCADS defined by (1.2.59) and (1.2.60) are said to be in Generalized Consensus with respect to a Transformation $H : \mathbb{R}^{m\times M\times N} \times \mathbb{R}^+ \to \mathbb{R}^{m\times M\times N}$ (see Fig. 1.2.4), abbreviated as GC, if for all $\epsilon > 0$, there exists an open subset $B_\epsilon \subset \mathbb{R}^{n\times M\times N} \times \mathbb{R}^{m\times M\times N}$, such that for any trajectory $(\mathbf{X}(t), \mathbf{Y}(t))$ of systems (1.2.59) and (1.2.60) with initial condition $(\mathbf{X}(0), \mathbf{Y}(0)) \in B_\epsilon$, the following property holds:

$$\|H(\mathbf{X}_m(t), t) - \mathbf{Y}(t)\| < \epsilon, \ \forall t \geq 0 \tag{2.2.7}$$

where

$$\begin{aligned}\mathbf{X}_m(t) &= (x_{l\,i,j}(t))_{m\times M\times N},\\ \mathbf{Y}(t) &= (y_{l\,i,j}(t))_{m\times M\times N}.\end{aligned}$$

Remark 2.2.3. As a special case, the transformation $H : \mathbb{R}^{m\times M\times N} \times \mathbb{R}^{+} \to \mathbb{R}^{m\times M\times N}$ may be independent of the time variable $t \in [0,\infty)$.

(4) GC Non-autonomous Bidirectional CCADS

Definition 2.8. The non-autonomous bidirectional CCADS defined by (1.2.68) and (1.2.69) are said to be in Generalized Consensus with respect to a Transformation $H : \mathbb{R}^{m\times M\times N} \times \mathbb{R}^{+} \to \mathbb{R}^{m\times M\times N}$ (see Fig. 1.2.4), if for all $\epsilon > 0$, there exists an open subset $B_\epsilon \subset \mathbb{R}^{n\times M\times N} \times \mathbb{R}^{m\times M\times N}$, such that for any trajectory $(\mathbf{X}(t), \mathbf{Y}(t))$ of systems (1.2.68) and (1.2.69) with initial condition $(\mathbf{X}(0), \mathbf{Y}(0)) \in B_\epsilon$, the following property holds:

$$\|H(\mathbf{X}_m(t), t) - \mathbf{Y}(t)\| < \epsilon,\ \forall t \geq 0 \tag{2.2.8}$$

where

$$\mathbf{X}_m(t) = (x_{l\,i,j}(t))_{m\times M\times N},$$
$$\mathbf{Y}(t) = (y_{l\,i,j}(t))_{m\times M\times N}.$$

Remark 2.2.4. As a special case, the transformation $H : \mathbb{R}^{m\times M\times N} \times \mathbb{R}^{+} \to \mathbb{R}^{m\times M\times N}$ may be independent of the time variable $t \in [0,\infty)$.

2.2.3 *PGC of Non-autonomous Bidirectional CDADS*

Definition 2.9. The non-autonomous bidirectional CDADS defined by (1.2.77) and (1.2.78) are said to be in Partial Generalized Consensus with respect to a Transformation $H : \mathbb{R}^{m\times M\times N} \times \mathbb{Z}^{+} \to \mathbb{R}^{m\times M\times N}$ (see Fig. 1.2.5), abbreviated as PGC, if for all $\epsilon > 0$, there exists an open subset $B_\epsilon \subset \mathbb{R}^{N_1\times M\times N} \times \mathbb{R}^{N_2\times M\times N}$, such that for any trajectory $(\mathbf{X}(k), \mathbf{Y}(k))$ of systems (1.2.77) and (1.2.78) with initial condition $(\mathbf{X}(0), \mathbf{Y}(0)) \in B_\epsilon$, the following property holds:

$$\|H(\mathbf{X}_m(k), k) - \mathbf{Y}_m(k)\| < \epsilon,\ \forall k \geq 0, \tag{2.2.9}$$

where

$$\mathbf{X}_m = (x_{l\,i,j})_{m\times M\times N},$$
$$\mathbf{Y}_m = (y_{l\,i,j})_{m\times M\times N}.$$

2.2.4 *PGC of Non-autonomous Bidirectional CCADS*

Definition 2.10. The bidirectional non-autonomous CCADS defined by (1.2.86) and (1.2.87) are said to be in Partial Generalized Consensus with respect to a Transformation $H : \mathbb{R}^{m\times M\times N} \times \mathbb{R}^{+} \to \mathbb{R}^{m\times M\times N}$ (see Fig. 1.2.2), abbreviated as PGC, if for all $\epsilon > 0$, there exists an open subset $B_\epsilon \subset \mathbb{R}^{N_1\times M\times N} \times \mathbb{R}^{N_2\times M\times N}$ such that for any trajectory $(\mathbf{X}(t), \mathbf{Y}(t))$ of systems (1.2.86) and (1.2.87) with initial condition $(\boldsymbol{X}(0), \mathbf{Y}(0)) \in B_\epsilon$, the following property holds:

$$\|H(\boldsymbol{X}_m(t), t) - \boldsymbol{Y}_m(t)\| < \epsilon, \ \forall t \geq 0 \tag{2.2.10}$$

where

$$\boldsymbol{X}_m(t) = (x_{l\,i,j}(t))_{m\times M\times N},$$
$$\boldsymbol{Y}_m(t) = (y_{l\,i,j}(t))_{m\times M\times N}.$$

Remark 2.2.5. As a special case, the transformation $H : \mathbb{R}^{m\times M\times N} \times \mathbb{R}^{+} \to \mathbb{R}^{m\times M\times N}$ may be independent of the time variable $t \in [0, \infty)$.

2.3 GC Theorems for Discrete and Continuous Arrays of Dynamic Systems

A basic GC problem on (discrete) arrays of difference equations (see systems (1.2.1) and (1.2.2), (1.2.10) and (1.2.11), (1.2.19) and (1.2.20), (1.2.28) and (1.2.29)) and (continuous) arrays of differentiable equations (see systems (1.2.37) and (1.2.38), (1.2.46) and (1.2.47), (1.2.55) and (1.2.56), (1.2.64) and (1.2.65)) is: if the coupled arrays can achieve GC with respect to a transformation H, then what should the general form of the coupled arrays be?

In the following subsections, ten general GC theorems on coupled arrays will be established. These theorems partially answer the above question for different arrays of dynamic systems.

2.3.1 *GC Theorem for Discrete Arrays*

Similarly to the GC theorem for discrete vector systems ([Wang *et al.* (2015a)]), the following GC theorem for CDADS can be established.

Theorem 2.1. *Let* $\mathbf{X}(k), \mathbf{Y}(k), F(\mathbf{X}(k))$ *and* $G(\mathbf{Y}(k), \mathbf{X}(k))$ *be defined by (1.2.3), (1.2.4), (1.2.7) and (1.2.8), respectively. Suppose that* $H : B \subset$

$\mathbb{R}^{m\times M\times N} \to \mathbb{R}^{m\times M\times N}$ is as defined in Definition 2.1. Then, the two systems (1.2.5) and (1.2.6) are in GC with respect to the transformation $\mathbf{Y}(k) = H(\mathbf{X}_m(k))$ if and only if the function $G(\mathbf{Y}(k),\mathbf{X}(k))$ is in the following form:

$$G(\mathbf{Y}(k),\mathbf{X}(k)) = H(F_m(\mathbf{X}(k))) - Q(\mathbf{X}(k),\mathbf{Y}(k)), \tag{2.3.1}$$

where

$$\begin{aligned} F_m(\mathbf{X}(k)) = ((&f_{1\,i,j}(\mathbf{X}(k)))_{M\times N},\dots,\\ &(f_{m\,i,j}(\mathbf{X}(k)))_{M\times N})^{\mathrm{T}}, \end{aligned} \tag{2.3.2}$$

and the function

$$\begin{aligned} Q(\mathbf{X}(k),\mathbf{Y}(k)) = ((&q_{1\,i,j}(\mathbf{X}(k),\mathbf{Y}(k)))_{M\times N},\dots,\\ &(q_{m\,i,j}(\mathbf{X}(k),\mathbf{Y}(k)))_{M\times N})^{\mathrm{T}} \end{aligned}$$

makes the zero solution of the following error equation (2.3.3) be stable on the open set B defined as in Definition 2.1:

$$\begin{aligned} \mathbf{e}(k+1) &= H(\mathbf{X}_m(k+1)) - \mathbf{Y}(k+1)\\ &= Q(\mathbf{X}(k),\mathbf{Y}(k)). \end{aligned} \tag{2.3.3}$$

Proof. Denote

$$\begin{aligned} &G(\mathbf{Y}(k),\mathbf{X}(k)) - H(F_m(\mathbf{X}(k)))\\ &\quad = -Q(\mathbf{X}(k),\mathbf{Y}(k)). \end{aligned}$$

Then,

$$\begin{aligned} \mathbf{e}(k+1) &= H(\mathbf{X}_m(k+1)) - \mathbf{Y}(k+1)\\ &= Q(\mathbf{X}(k),\mathbf{Y}(k)). \end{aligned} \tag{2.3.4}$$

Therefore, the two dynamic systems (1.2.5) and (1.2.6) are in GC with respect to the transform H if and only if the function $Q(\mathbf{X}_m(k),\mathbf{Y}(k))$ makes the zero solution of the error equation (2.3.4) be stable on the set B defined as in Definition 2.1.

This completes the proof. □

Remark 2.3.1. In fact, $Q(\mathbf{X}(k),\mathbf{Y}(k))$ can be easily constructed. For example, take $Q(\mathbf{X}(k),\mathbf{Y}(k)) = A\mathbf{e}(k)$, where A is a matrix whose eigenvalues $\lambda's$ have the following properties:

(1) $|\lambda| < 1$, or

(2) simple eigenvalue $|\lambda| = 1$.

2.3.2 GC Theorem for Discrete Bidirectional Arrays

Similarly to the GS theorem for discrete bidirectional vector systems [Yang *et al.* (2015)], the following GC theorem for discrete bidirectional can be established.

Theorem 2.2. *Let* $\mathbf{X}(k)$, $\mathbf{Y}(k)$, $F(\mathbf{X}(k), \mathbf{Y}(k))$ *and* $G(\mathbf{Y}(k), \mathbf{X}(k))$ *be defined by (1.2.12), (1.2.13), (1.2.16) and (1.2.17), respectively. Suppose that the transformation* $H : \mathbb{R}^{m\times M\times N} \to \mathbb{R}^{m\times M\times N}$ *is defined as in Definition 2.2. Then, the bidirectional CDADS defined by (1.2.14) and (1.2.15) are in GC with respect to the transformation* $\mathbf{Y}(k) = H(\mathbf{X}_m(k))$ *if, and only if, the function* $G(\mathbf{Y}(k), \mathbf{X}(k))$ *is in the following form:*

$$\begin{aligned} G(\mathbf{Y}(k), \mathbf{X}(k)) = H(F_m(\mathbf{X}(k), \mathbf{Y}(k))) \\ -Q(\mathbf{X}(k), \mathbf{Y}(k)), \end{aligned} \tag{2.3.5}$$

where the function

$$\begin{aligned} Q(\mathbf{X}(k), \mathbf{Y}(k)) = ((q_{1\,i,j}(\mathbf{X}(k), \mathbf{Y}(k)))_{M\times N}, \\ (q_{2\,i,j}(\mathbf{X}(k), \mathbf{Y}(k)))_{M\times N}, \\ \ldots, (q_{m\,i,j}(\mathbf{X}(k), \mathbf{Y}(k)))_{M\times N})^{\mathrm{T}} \end{aligned}$$

makes the zero solution of the following error equation be stable on the open set B *defined as in Definition 2.2:*

$$\begin{aligned} \mathbf{e}(k+1) &= H(\mathbf{X}(k+1)) - \mathbf{Y}(k+1) \\ &= Q(\mathbf{X}(k), \mathbf{Y}(k)). \end{aligned} \tag{2.3.6}$$

Proof. Denote

$$\begin{aligned} &G(\mathbf{Y}(k), \mathbf{X}(k)) - H(F_m(\mathbf{X}(k), \mathbf{Y}(k))) \\ &= -Q(\mathbf{X}(k), \mathbf{Y}(k)). \end{aligned}$$

Then,

$$\begin{aligned} \mathbf{e}(k+1) &= H(\mathbf{X}_m(k+1)) - \mathbf{Y}(k+1) \\ &= Q(\mathbf{X}(k), \mathbf{Y}(k)). \end{aligned} \tag{2.3.7}$$

Therefore, the two dynamic systems (1.2.14) and (1.2.15) are in GC with respect to the transform H if, and only if, the function $Q(\mathbf{X}_m(k), \mathbf{Y}(k))$ makes the zero solution of the error equation (2.3.7) be stable on the set B defined by Definition 2.2.

This completes the proof. □

2.3.3 *GC Theorem for Discrete Non-autonomous Arrays*

Similarly to the case of discrete non-autonomous vector difference systems [Zhang *et al.* (2015b)], the following result can be established.

Theorem 2.3. *Let* $\mathbf{X}(k), \mathbf{Y}(k), F(\mathbf{X}(k), k)$ *and* $G(\mathbf{Y}(k), \mathbf{X}(k), k)$ *be defined by (1.2.21), (1.2.22), (1.2.25), and (1.2.26), respectively. Suppose that* $H : \mathbb{R}^{m\times M\times N} \times \mathbb{Z}^{+} \to \mathbb{R}^{m\times M\times N}$ *be defined by Definition 2.3. Then, the non-autonomous CDADS defined by (1.2.23) and (1.2.24) are in GC with respect to the transformation* $\mathbf{Y}(k) = H(\mathbf{X}_m(k))$ *if, and only if, the function* $G(\mathbf{Y}(k), \mathbf{X}(k), k)$ *is in the following form:*

$$\begin{aligned} G(\mathbf{Y}(k), \mathbf{X}(k), k) = H(&F_m(\mathbf{X}(k), k), k+1) \\ &-Q(\mathbf{X}(k), \mathbf{Y}(k), k) \end{aligned} \tag{2.3.8}$$

where the function

$$\begin{aligned} Q(\mathbf{X}(k), \mathbf{Y}(k), k) = (&(q_{1\,i,j}(\mathbf{X}(k), \mathbf{Y}(k), k))_{M\times N}, \\ &(q_{2\,i,j}(\mathbf{X}(k), \mathbf{Y}(k), k))_{M\times N}, \ldots, \\ &(q_{m\,i,j}(\mathbf{X}(k), \mathbf{Y}(k), k))_{M\times N})^{\mathrm{T}} \end{aligned}$$

makes the zero solution of the following error equation be stable on the open set B defined as in Definition 2.3:

$$\begin{aligned} \mathbf{e}(k+1) &= H(\mathbf{X}_m(k+1), k+1) - \mathbf{Y}(k+1) \\ &= Q(\mathbf{X}(k), \mathbf{Y}(k), k). \end{aligned} \tag{2.3.9}$$

Proof. Denote

$$\begin{aligned} &G(\mathbf{Y}(k), \mathbf{X}(k), k) - H(F_m(\mathbf{X}(k)), k+1) \\ &= -Q(\mathbf{X}(k), \mathbf{Y}(k), k). \end{aligned}$$

Then,

$$\begin{aligned} \mathbf{e}(k+1) &= H(\mathbf{X}_m(k+1), k+1) - \mathbf{Y}(k+1) \\ &= Q(\mathbf{X}(k), \mathbf{Y}(k), k). \end{aligned} \tag{2.3.10}$$

Therefore, the two dynamic systems (1.2.23) and (1.2.24) are in GC with respect to the transform H if, and only if, the function $Q(\mathbf{X}(k), \mathbf{Y}(k), k)$ makes the zero solution of the error equation (2.3.10) be stable on the set B defined by Definition 2.3.

This completes the proof. □

2.3.4 GC Theorem for Discrete Non-autonomous Bidirectional Arrays

Similarly to the case of the GS theorem for discrete non-autonomous vector difference systems [Liu *et al.* (2010)], the following result can be established.

Theorem 2.4. *Let* $\mathbf{X}(k)$, $\mathbf{Y}(k)$, $F(\mathbf{X}(k), \mathbf{Y}(k), k)$ *and* $G(\mathbf{Y}(k), \mathbf{X}(k), k)$ *be defined by (1.2.30), (1.2.31), (1.2.34), (1.2.35), respectively. Suppose that* $H : \mathbb{R}^{m\times M\times N} \times \mathbb{Z}^+ \to \mathbb{R}^{m\times M\times N}$ *is defined by Definition 1.11. Then, the non-autonomous bidirectional CDADSs defined by (1.2.32) and (1.2.33) are in GC with respect to the transformation* $\mathbf{Y}(k) = H(\mathbf{X}_m(k), k)$ *if, and only if, the function* $G(\mathbf{Y}(k), \mathbf{X}(k), k)$ *is in the following form:*

$$G(\mathbf{Y}(k), \mathbf{X}(k), k) = H(F_m(\mathbf{X}(k), \mathbf{Y}(k), k), k+1) - Q(\mathbf{X}(k), \mathbf{Y}(k), k) \tag{2.3.11}$$

where the function

$$Q(\mathbf{X}(k), \mathbf{Y}(k), k) = ((q_{1\,i,j}(\mathbf{X}(k), \mathbf{Y}(k), k))_{M\times N}, (q_{2\,i,j}(\mathbf{X}(k), \mathbf{Y}(k), k))_{M\times N}, \dots, (q_{m\,i,j}(\mathbf{X}(k), \mathbf{Y}(k), k))_{M\times N})^T$$

makes the zero solution of the following error equation be stable on the open set B *defined as in Definition 2.4:*

$$\begin{aligned}\mathbf{e}(k+1) &= H(\mathbf{X}_m(k+1), k+1) - \mathbf{Y}(k+1)\\ &= Q(\mathbf{X}(k), \mathbf{Y}(k), k).\end{aligned} \tag{2.3.12}$$

Proof. Denote

$$G(\mathbf{Y}(k), \mathbf{X}(k), k) - H(F_m(\mathbf{X}(k), \mathbf{Y}(k), k), k+1) = -Q(\mathbf{X}(k), \mathbf{Y}(k), k).$$

Then,

$$\begin{aligned}\mathbf{e}(k+1) &= H(\mathbf{X}_m(k+1), k+1) - \mathbf{Y}(k+1)\\ &= H(F_m(\mathbf{X}(k), \mathbf{Y}(k), k), k+1) - H(F_m(\mathbf{X}(k), \mathbf{Y}(k), k+1)\\ &\quad +Q(\mathbf{X}, \mathbf{Y}, k)\\ &= Q(\mathbf{X}, \mathbf{Y}, k).\end{aligned}$$

Therefore, two dynamic systems (1.2.32) and (1.2.33) are in GC with respect to the transformation H if, and only if, the function $Q(\mathbf{X}(k), \mathbf{Y}(k), k)$ makes the zero solution of the error equation (2.3.12) be stable on the set B defined as in Definition 2.4.

This completes the proof. □

2.3.5 *GC Theorem for an Array of Differential Systems*

Similarly to the GS Theorem 1.5 for an array of differential equations, the following corresponding result can be established.

Theorem 2.5. *Let* $\mathbf{X}(t)$, $\mathbf{Y}(t)$, $F(\mathbf{X}(t))$ *and* $G(\mathbf{Y}(t), \mathbf{X}(t))$ *be defined by (1.2.39), (1.2.40), (1.2.43), and (1.2.44), respectively. Then, the two systems (1.2.41) and (1.2.42) are in GC with respect to the transformation* $\mathbf{Y} = H(\mathbf{X}_m)$ *defined as in Definition 2.5 (see Fig.1.3.1, also and Fig. 7 in* [Min and Chen (2013)]*) if and only if the function* $G(\mathbf{Y}(t), \mathbf{X}(t))$ *is in the following form:*

$$\begin{aligned} g_{l\,i,j}(\mathbf{Y}(t), \mathbf{X}(t)) = \sum_{l'=1}^{m}\sum_{i'=1}^{M}\sum_{j'=1}^{N} & \frac{\partial h_{l\,i,j}(\mathbf{X}_m(t))}{\partial x_{l'\,i',j'}} \\ & \times f_{l'\,i',j'}(\mathbf{X}(t)) - q_{l\,i,j}(\mathbf{X}(t), \mathbf{Y}(t)), \qquad (2.3.13) \\ & l = 1, \cdots, m, i = 1, \cdots, M, \\ & j = 1, \cdots, N, \end{aligned}$$

where the function $q_{l\,i,j}(\mathbf{X}(t), \mathbf{Y}(t))$ *make the zero solution of the error equation*

$$\begin{aligned} \frac{d\mathbf{e}}{dt} &= \frac{d(H(\mathbf{X}_m(t)) - \mathbf{Y}\,(t))}{dt} \\ &= (q_{l\,i,j}(\mathbf{X}, \mathbf{Y}, t))_{m\times M\times N} \end{aligned} \qquad (2.3.14)$$

be stable on the open set B *defined as in Definition 2.5.*

Proof. $\Rightarrow$ Since H is continuously differentiable, one can write out $g_{l\,i,j}(\mathbf{Y}, \mathbf{X})$ as

$$\begin{aligned} g_{l\,i,j}(\mathbf{Y}, \mathbf{X}) = \sum_{l'=1}^{m}\sum_{i'=1}^{M}\sum_{j'=1}^{N} & \frac{\partial h_{l\,i,j}(\mathbf{X}_m)}{\partial x_{l'\,i',j'}} f_{l'\,i',j'}(\mathbf{X}, \mathbf{Y}) \\ & - q_{l\,i,j}(\mathbf{X}, \mathbf{Y}). \end{aligned}$$

Let

$$\mathbf{e} = H(\mathbf{X}_m) - \mathbf{Y}.$$

Then, the error equation has the following form:

$$\begin{aligned}
\frac{d\mathbf{e}}{dt} &= \left(\frac{dh_{l\,i,j}(\mathbf{X}_m)}{dt}\right)_{m\times M\times N} - \frac{d\mathbf{Y}}{dt} \\
&= \left(\sum_{l'=1}^{m}\sum_{i'=1}^{M}\sum_{j'=1}^{N} \frac{\partial h_{l\,i,j}(\mathbf{X}_m)}{\partial x_{l'\,i',j'}}\dot{x}_{l'\,i',j'}\right)_{m\times M\times N} \\
&\quad - (g_{l\,i,j}(\mathbf{Y},\mathbf{X}))_{m\times M\times N} \\
&= \left(\sum_{l'=1}^{m}\sum_{i'=1}^{M}\sum_{j'=1}^{N} \frac{\partial h_{l\,i,j}(\mathbf{X}_m)}{\partial x_{l'\,i',j'}} f_{l'\,i',j'}(\mathbf{X},\mathbf{Y})\right)_{m\times M\times N} \\
&\quad - (g_{l\,i,j}(\mathbf{Y},\mathbf{X}))_{m\times M\times N} \\
&= (q_{l\,i,j}(\mathbf{X},\mathbf{Y}))_{m\times M\times N}\,.
\end{aligned} \tag{2.3.15}$$

By assumption, CCADS (1.2.41) and (1.2.42) are in GC with respect to the transformation H. Therefore, $q_{l\,i,j}(\mathbf{X},\mathbf{Y})$ ensures the zero solution of system (2.3.15) be stable for any initial condition in the set B defined as in Definition 2.5.

$\Leftarrow$ If the function $G_{l\,i,j}(\mathbf{Y},\mathbf{X})$ can be represented by equation (2.3.13), and $(q_{l\,i,j}(\mathbf{X},\mathbf{Y}))_{m\times M\times N}$ makes the zero solution of equation (2.3.14) be stable on the set B defined by Definition 2.5, then CCADS (1.2.41) and (1.2.42) are in GC with respect to the transformation $\mathbf{Y} = H(\mathbf{X}_m)$.

This completes the proof. $\square$

2.3.6 *GC Theorem for a Bidirectional Array of Differential Systems*

The GC theorem for a coupled bidirectional vector system of differential equations [Yang *et al.* (2015)] is now extended to the case of an array of differential equations:

Theorem 2.6. *Suppose that* $H : \mathbb{R}^{m\times M\times N} \to \mathbb{R}^{m\times M\times N}$ *defined as in Definition 2.6 (see also Fig. 1.2.2) is continuously differentiable. Let* $\mathbf{X}(t)$, $\mathbf{Y}(t)$, $F(\mathbf{X}(t),\mathbf{Y}(t))$ *and* $G(\mathbf{Y}(t),\mathbf{X}(t))$ *be defined by (1.2.48), (1.2.49), (1.2.52) and (1.2.53), respectively. Then, the bidirectional CCADS defined by (1.2.50) and (1.2.51) are in GC with respect to the transformation*

$\mathbf{Y} = H(\mathbf{X}_m)$ if, and only if, the function $G(\mathbf{Y}, \mathbf{X})$ is in the following form:

$$\begin{aligned} g_{l\,i,j}(\mathbf{Y}, \mathbf{X}) = & \sum_{l'=1}^{m}\sum_{i'=1}^{M}\sum_{j'=1}^{N} \frac{\partial h_{l\,i,j}(\mathbf{X}_m)}{\partial x_{l'\,i',j'}} f_{l'\,i',j'}(\mathbf{X}, \mathbf{Y}) \\ & -q_{l\,i,j}(\mathbf{X}, \mathbf{Y}), \\ & l = 1, \cdots, m, i = 1, \cdots, M, \\ & j = 1, \cdots, N, \end{aligned} \tag{2.3.16}$$

where the function $q_{l\,i,j}(\mathbf{X}, \mathbf{Y})$ make the zero solution of the error equation

$$\begin{aligned} \frac{d\mathbf{e}}{dt} &= \frac{d(H(\mathbf{X}_m) - \mathbf{Y})}{dt} \\ &= (q_{l\,i,j}(\mathbf{X}, \mathbf{Y}))_{m\times M\times N} \end{aligned} \tag{2.3.17}$$

be stable on the open set B defined as in Definition 1.15.

Proof. $\Rightarrow$ Since H is continuously differentiable, one can write

$$\begin{aligned} g_{l\,i,j}(\mathbf{Y}, \mathbf{X}) = & \sum_{l'=1}^{m}\sum_{i'=1}^{M}\sum_{j'=1}^{N} \frac{\partial h_{l\,i,j}(\mathbf{X}_m)}{\partial x_{l'\,i',j'}} f_{l'\,i',j'}(\mathbf{X}, \mathbf{Y}) \\ & -q_{l\,i,j}(\mathbf{X}, \mathbf{Y}). \end{aligned}$$

Let

$$\mathbf{e} = H(\mathbf{X}_m) - \mathbf{Y}.$$

Then, the error equation has the following form:

$$\begin{aligned} \frac{d\mathbf{e}}{dt} &= \left(\frac{dh_{l\,i,j}(\mathbf{X}_m)}{dt}\right)_{m\times M\times N} - \frac{d\mathbf{Y}}{dt} \\ &= \left(\sum_{l'=1}^{m}\sum_{i'=1}^{M}\sum_{j'=1}^{N} \frac{\partial h_{l\,i,j}(\mathbf{X}_m)}{\partial x_{l'\,i',j'}} \dot{x}_{l'\,i',j'}\right)_{m\times M\times N} \\ &\quad - (g_{l\,i,j}(\mathbf{Y}, \mathbf{X}))_{m\times M\times N} \\ &= \left(\sum_{l'=1}^{m}\sum_{i'=1}^{M}\sum_{j'=1}^{N} \frac{\partial h_{l\,i,j}(\mathbf{X}_m)}{\partial x_{l'\,i',j'}} f_{l'\,i',j'}(\mathbf{X}, \mathbf{Y})\right)_{m\times M\times N} \\ &\quad - (g_{l\,i,j}(\mathbf{Y}, \mathbf{X}))_{m\times M\times N} \\ &= (q_{l\,i,j}(\mathbf{X}, \mathbf{Y}))_{m\times M\times N}. \end{aligned} \tag{2.3.18}$$

By assumption, CCADS (1.2.50) and (1.2.51) are in GC with respect to the transformation H. Therefore, $q_{l\,i,j}(\mathbf{X}, \mathbf{Y})$ ensures the zero solution of system (2.3.18) to be stable for any initial condition in the set B defined as in Definition 2.6.

$\Leftarrow$ If the function $g_{l\,i,j}(\boldsymbol{Y}, \boldsymbol{X})$ can be represented by equation (2.3.16), and $(q_{l\,i,j}(\mathbf{X}, \boldsymbol{Y}))_{m\times M\times N}$ makes the zero solution of equation (2.3.17) be stable on the set B defined by Definition 2.6, then CCADS (1.2.50) and (1.2.51) are in GC with respect to the transformation $\boldsymbol{Y} = H(\boldsymbol{X}_m)$.

This completes the proof. □

2.3.7 GC Theorem for a Non-autonomous Array of Differential Systems

This subsection extends the GC theorem for a non-autonomous coupled vector system of differential equations [Zhang *et al.* (2015a)] to the case of an array of differential equations.

Theorem 2.7. *Suppose that $H : \mathbb{R}^{m\times M\times N} \times \mathbb{R}^{+} \to \mathbb{R}^{m\times M\times N}$ defended by Definition 2.7 is continuously differentiable. Let $\mathbf{X}(t), \mathbf{Y}(t), F(\mathbf{X}(t), t)$ and $G(\mathbf{Y}(t), \mathbf{X}(t), t)$ be defined by (1.2.57), (1.2.58), (1.2.61), (1.2.62), respectively. Then, the non-autonomous CCADS defined by (1.2.59) and (1.2.60) are in GC with respect to the transformation $\mathbf{Y} = H(\mathbf{X}_m, t)$ if and only if the function $G(\mathbf{Y}, \mathbf{X}, t)$ is in the following form:*

$$\begin{aligned} g_{l\,i,j}(\mathbf{Y}, \mathbf{X}, t) = & \sum_{l'=1}^{m}\sum_{i'=1}^{M}\sum_{j'=1}^{N} \frac{\partial h_{l\,i,j}(\mathbf{X}_m, t)}{\partial x_{l'\,i',j'}} f_{l'\,i',j'}(\mathbf{X}, t) \\ & + \frac{\partial h_{l\,i,j}(\mathbf{X}_m, t)}{\partial t} - q_{l\,i,j}(\mathbf{X}, \mathbf{Y}, t), \\ & l = 1, \cdots, m, i = 1, \cdots, M, \\ & j = 1, \cdots, N, \end{aligned} \tag{2.3.19}$$

where the function $q_{l\,i,j}(\mathbf{X}, \mathbf{Y}, t)$ make the error solution of the error equation

$$\begin{aligned} \frac{d\mathbf{e}}{dt} &= \frac{d(H(\mathbf{X}_m, t) - \mathbf{Y})}{dt} \\ &= (q_{l\,i,j}(\mathbf{X}, \mathbf{Y}, t))_{m\times M\times N} \end{aligned} \tag{2.3.20}$$

be stable on the open set B defined as in Definition 2.7.

Proof. $\Rightarrow$ Since $H(\mathbf{X}_m, t)$ is continuously differentiable, one can write

$$g_{l\,i,j}(\mathbf{Y}, \mathbf{X}, t) = \sum_{l'=1}^{m}\sum_{i'=1}^{M}\sum_{j'=1}^{N} \frac{\partial h_{l\,i,j}(\mathbf{X}_m, t)}{\partial x_{l'\,i',j'}} f_{l'\,i',j'}(\mathbf{X}, t) + \frac{\partial h_{l\,i,j}(\mathbf{X}_m, t)}{\partial t} - q_{l\,i,j}(\mathbf{X}, \mathbf{Y}, t). \tag{2.3.21}$$

Let

$$\mathbf{e} = H(\mathbf{X}_m, t) - \mathbf{Y}.$$

Then, the error equation has the following form:

$$\begin{aligned} \frac{d\mathbf{e}}{dt} &= \left(\frac{dh_{l\,i,j}(\mathbf{X}_m, t)}{dt}\right)_{m\times M\times N} - \frac{d\mathbf{Y}}{dt} \\ &= \left(\sum_{l'=1}^{m}\sum_{i'=1}^{M}\sum_{j'=1}^{N} \frac{\partial h_{l\,i,j}(\mathbf{X}_m, t)}{\partial x_{l'\,i',j'}} \dot{x}_{l'\,i',j'}\right)_{m\times M\times N} \\ &\quad + \left(\frac{\partial h_{l\,i,j}(\mathbf{X}_m, t)}{\partial t}\right)_{m\times M\times N} \\ &\quad - (g_{l\,i,j}(\mathbf{Y}, \mathbf{X}, t))_{m\times M\times N} \\ &= \left(\sum_{l'=1}^{m}\sum_{i'=1}^{M}\sum_{j'=1}^{N} \frac{\partial h_{l\,i,j}(\mathbf{X}_m, t)}{\partial x_{l'\,i',j'}} f_{l'\,i',j'}(\mathbf{X}, t)\right)_{m\times M\times N} \\ &\quad + \left(\frac{\partial h_{l\,i,j}(\mathbf{X}_m, t)}{\partial t}\right)_{m\times M\times N} \\ &\quad - (g_{l\,i,j}(\mathbf{Y}, \mathbf{X}, t))_{m\times M\times N} \\ &= (q_{l\,i,j}(\mathbf{X}, \mathbf{Y}, t))_{m\times M\times N}. \end{aligned} \tag{2.3.22}$$

By assumption, the two CCADS (1.2.59) and (1.2.60) are in GC with respect to the transformation H. Therefore, $q_{l\,i,j}(\mathbf{X}, \mathbf{Y}, t)$ ensures the zero solution of the error equation (2.3.20) be stable for any initial condition in the set B defined as in Definition 2.7.

$\Leftarrow$ If the function $g_{l\,i,j}(\mathbf{Y}, \mathbf{X}, t)$ can be represented by equation (2.3.19), and $(q_{l\,i,j}(\mathbf{X}, \mathbf{Y}, t))_{m\times M\times N}$ makes the zero solution of the error equation (2.3.20) be stable on the set B defined as in Definition 2.7, then the two CCADS (1.2.59) and (1.2.60) are in GC with respect to the transformation $H(\mathbf{X}_m, t) = \mathbf{Y}$.

This completes the proof. □

2.3.8 GC Theorem for a Non-autonomous Bidirectional Array of Differential Systems

Similarly to GS Theorem 1.9 for a non-autonomous bidirectional coupled system of differential equations (also see Theorem 9 given in [Min and Chen (2013)]), the following result can be established.

Theorem 2.8. *Suppose that $H : \mathbb{R}^{m\times M\times N} \times \mathbb{R}^+ \to \mathbb{R}^{m\times M\times N}$ defined by Definition 2.8 is continuously differentiable. Let $\mathbf{X}(t)$, $\mathbf{Y}(t)$, $F(\mathbf{X}(t), \mathbf{Y}(t), t)$ and $G(\mathbf{Y}(t), \mathbf{X}(t), t)$ be defined by (1.2.66), (1.2.67), (1.2.70) and (1.2.71), respectively. Then, the non-autonomous bidirectional CCADS (1.2.68) and (1.2.69) are in GC with respect to the transformation $\mathbf{Y} = H(\mathbf{X}_m, t)$ if and only if the function $G(\mathbf{Y}, \mathbf{X}, t)$ is in the following form:*

$$\begin{aligned} g_{l\,i,j}(\mathbf{Y}, \mathbf{X}, t) = & \sum_{l'=1}^{n}\sum_{i'=1}^{M}\sum_{j'=1}^{N} \frac{\partial h_{l\,i,j}(\mathbf{X}_m, t)}{\partial x_{l'\,i',j'}} \\ & \times f_{l'\,i',j'}(\mathbf{X}, \mathbf{Y}, t) + \frac{\partial h_{l\,i,j}(\mathbf{X}_m, t)}{\partial t} \\ & - q_{l\,i,j}(\mathbf{X}, \mathbf{Y}, t), \\ & l = 1, \cdots, n, i = 1, \cdots, M, \\ & j = 1, \cdots, N \end{aligned} \tag{2.3.23}$$

where the function $q_{l\,i,j}(\mathbf{X}, \mathbf{Y}, t)$ make the zero solution of the error equation

$$\begin{aligned} \frac{d\mathbf{e}}{dt} &= \frac{d(H(\mathbf{X}_m, t) - \mathbf{Y})}{dt} \\ &= (q_{l\,i,j}(\mathbf{X}, \mathbf{Y}, t))_{m\times M\times N} \end{aligned} \tag{2.3.24}$$

be stable on the open set B defined as in Definition 2.8.

Proof. $\Rightarrow$ Since $H(\mathbf{X}_m, t)$ is continuously differentiable, one can write

$$\begin{aligned} g_{l\,i,j}(\mathbf{Y}, \mathbf{X}, t) = & \sum_{l'=1}^{n}\sum_{i'=1}^{M}\sum_{j'=1}^{N} \frac{\partial h_{l\,i,j}(\mathbf{X}_m)}{\partial x_{l'\,i',j'}} \\ & \times f_{l'\,i',j'}(\mathbf{X}, \mathbf{Y}, t) + \frac{\partial h_{l\,i,j}(\mathbf{X}_m, t)}{\partial t} \\ & - q_{l\,i,j}(\mathbf{X}, \mathbf{Y}, t). \end{aligned}$$

Let

$$\mathbf{e} = H(\mathbf{X}_m, t) - \mathbf{Y}.$$

Then, the error equation has the following form:

$$\begin{aligned}
\frac{d\boldsymbol{e}}{dt} &= \left(\frac{dh_{l\,i,j}(\mathbf{X}_m,t)}{dt}\right)_{m\times M\times N} - \frac{d\mathbf{Y}}{dt} \\
&= \left(\sum_{l'=1}^{m}\sum_{i'=1}^{M}\sum_{j'=1}^{N}\frac{\partial h_{l\,i,j}(\mathbf{X}_m)}{\partial x_{l'\,i',j'}}\dot{x}_{l'\,i',j'}\right)_{m\times M\times N} \\
&\quad + \left(\frac{\partial h_{l\,i,j}(\mathbf{X}_m,t)}{\partial t}\right)_{m\times M\times N} \\
&\quad - (g_{l\,i,j}(\mathbf{Y},\mathbf{X},t))_{m\times M\times N} \\
&= \left(\sum_{l'=1}^{m}\sum_{i'=1}^{M}\sum_{j'=1}^{N}\frac{\partial h_{l\,i,j}(\mathbf{X}_m)}{\partial x_{l'\,i',j'}}\right. \\
&\quad \left.\times f_{l'\,i',j'}(\mathbf{X},\mathbf{Y},t)\right)_{m\times M\times N} + \left(\frac{\partial h_{l\,i,j}(\mathbf{X}_m,t)}{\partial t}\right)_{m\times M\times N} \\
&\quad - (g_{l\,i,j}(\mathbf{Y},\mathbf{X},t))_{m\times M\times N} \\
&= (q_{l\,i,j}(\mathbf{X},\mathbf{Y},t))_{m\times M\times N}.
\end{aligned}$$

By assumption, CCADS (1.2.68) and (1.2.69) are in GS with respect to the transformations H. Therefore, $q_{l\,i,j}(\mathbf{X},\mathbf{Y},t)$ make the zero solution of the error equation (2.3.24) be stable for any initial condition in the set B defined as in Definition 2.8.

$\Leftarrow$ If the function $g_{l\,i,j}(\mathbf{Y},\mathbf{X},t)$ can be represented by equation (2.3.23), and $(q_{l\,i,j}(\mathbf{X},\mathbf{Y},t))_{m\times M\times N}$ makes the zero solution of equation (2.3.24) be stable on the set B defined by Definition 2.8, then condition (2.2.8) holds. This means the two CCADS (1.2.68) and (1.2.69) are in GC with respect to the transformation $H(\mathbf{X}_m,t) = \mathbf{Y}$.

This completes the proof. □

2.3.9 *PGC Theorem for Non-autonomous Bidirectional CDADS*

Non-autonomous bidirectional coupled discrete arrays are a kind of the most complex and general discrete networks. The partial GC (PGC) theory for them is of common interest.

Similarly to the partial GS Theorem 1.10 for non-autonomous bidirectional coupled discrete arrays (also see Theorem 10 given in [Min and Chen (2013)]), we have the following result.

Theorem 2.9. *Let* $\mathbf{X}(k)$, $\mathbf{Y}(k)$, $F(\mathbf{X}(k),\mathbf{Y}(k),k)$ *and* $G(\mathbf{Y}(k),\mathbf{X}(k),k)$ *be* *defined by (1.2.75), (1.2.76), (1.2.79), and (1.2.80), respectively. Suppose that the transformation* $H : \mathbb{R}^{m\times M\times N} \times \mathbb{Z}^+ \to \mathbb{R}^{m\times M\times N}$ *is defined by Definition 2.9. Then, the non-autonomous bidirectional CADDS defined by (1.2.77) and (1.2.78) are in partial GC (PGC) with respect to the transformation* $\mathbf{Y}_m(k) = H(\mathbf{X}_m(k), k)$ *if, and only if, the function*

$$G_m(\mathbf{Y}(k), \mathbf{X}(k), k) = (g_{l\,i,j}(\mathbf{Y}(k), \mathbf{X}(k), k))_{m\times M\times M} \tag{2.3.25}$$

given in (1.2.80) is in the following form:

$$\begin{aligned} G_m(\mathbf{Y}(k), \mathbf{X}(k), k) = H(F_m(\mathbf{X}(k), \mathbf{Y}(k), k), k+1) \\ -Q(\mathbf{X}(k), \mathbf{Y}(k), k), \end{aligned} \tag{2.3.26}$$

where the function

$$\begin{aligned} Q(\mathbf{X}(k), \mathbf{Y}(k), k) = [&(q_{1\,i,j}\mathbf{X}(k), \mathbf{Y}(k), k))_{M\times N}, \\ &(q_{2\,i,j}(\mathbf{X}(k), \mathbf{Y}(k), k))_{M\times N}, \ldots, \\ &(q_{m\,i,j}(\mathbf{X}(k), \mathbf{Y}(k), k))_{M\times N}]^{\mathrm{T}} \end{aligned} \tag{2.3.27}$$

makes the zero solution of the following error equation be stable on the open set B defined as in Definition 2.9:

$$\begin{aligned} \mathbf{e}(k+1) &= H(\mathbf{X}_m(k+1), k+1) - \mathbf{Y}_m(k+1) \\ &= Q(\mathbf{X}(k), \mathbf{Y}(k), k). \end{aligned} \tag{2.3.28}$$

Proof. Denote

$$\begin{aligned} &G_m(\mathbf{Y}(k), \mathbf{X}(k), k) - H(F_m(\mathbf{X}(k), \mathbf{Y}(k), k+1)) \\ &\quad = -Q(\mathbf{X}(k), \mathbf{Y}(k), k). \end{aligned} \tag{2.3.29}$$

Then,

$$\begin{aligned} \mathbf{e}(k+1) &= H(\mathbf{X}_m(k+1), k+1) - \mathbf{Y}_m(k+1). \\ &= Q(\mathbf{X}(k), \mathbf{Y}(k), k). \end{aligned} \tag{2.3.30}$$

Therefore, the two systems (1.2.77) and (1.2.78) are in PGC via the transformation H defined by Definition 2.9 if and only if the function $Q(\mathbf{X}(k), \mathbf{Y}(k), k)$ makes the zero solution of the error equation (2.3.28) be stable.

This completes the proof. □

Remark 2.3.2. In fact, $Q(\mathbf{X}(k), \mathbf{Y}(k), k)$ can be easily constructed. For example, take $Q(\mathbf{X}(k), \mathbf{Y}(k), k) = A\mathbf{e}(k)$, where A is a matrix whose eigenvalues $\lambda's$ have the following properties:

(1) $|\lambda| < 1$, or

(2) simple eigenvalue $|\lambda| = 1$.

2.3.10 *PGC Theorem for Non-autonomous Bidirectional CCADS*

Non-autonomous bidirectional arrays of continuously differentiable arrays are a kind of the most complex and general continuous networks. The corresponding partial GC (PGC) theory is of common interest.

Similarly to the PGS Theorem 1.11 for non-autonomous bidirectional coupled continuously differentiable systems (also see Theorem 11 given in [Min and Chen (2013)], the following result can be established.

Theorem 2.10. *Suppose that $H : \mathbb{R}^{m\times M\times N} \times \mathbb{R}^+ \to \mathbb{R}^{m\times M\times N}$ is continuously differentiable. Let $\mathbf{X}(t)$, $\mathbf{Y}(t)$, $F(\mathbf{X}(t), \mathbf{Y}(t), t)$ and $G(\mathbf{Y}(t), \mathbf{X}(t), t)$ be defined by (1.2.84), (1.2.85), (1.2.88), and (1.2.89), respectively. Then, the non-autonomous bidirectional CCADS defined by (1.2.86) and (1.2.87) are in PGC with respect to the transformation $\mathbf{Y}_m(t) = H(\mathbf{X}_m(t), t)$ defined by Definition 2.10 if and only if the function $G(\mathbf{Y}, \mathbf{X}, t)$ is in the following form:*

$$\begin{aligned} g_{l\,i,j}(\mathbf{Y}, \mathbf{X}, t) = & \sum_{l'=1}^{m}\sum_{i'=1}^{M}\sum_{j'=1}^{N} \frac{\partial h_{l\,i,j}(\mathbf{X}_m, t)}{\partial x_{l'\,i',j'}} \\ & \times f_{l'\,i',j'}(\mathbf{X}, \mathbf{Y}, t) + \frac{\partial h_{l\,i,j}(\mathbf{X}_m, t)}{\partial t} \\ & - q_{l\,i,j}(\mathbf{X}, \mathbf{Y}, t), \\ & l = 1, \cdots, m, i = 1, \cdots, M, \\ & j = 1, \cdots, N, \end{aligned} \tag{2.3.31}$$

where the function $q_{l\,i,j}(\mathbf{X}, \mathbf{Y}, t)$ make the zero solution of the error equation

$$\begin{aligned} \frac{d\mathbf{e}}{dt} &= \frac{d(H(\mathbf{X}_m, t) - \mathbf{Y}_m)}{dt} \\ &= (q_{l\,i,j}(\mathbf{X}, \mathbf{Y}, t))_{m\times M\times N} \end{aligned} \tag{2.3.32}$$

be stable on the open set B defined as in Definition 2.10.

Proof. $\Rightarrow$ Since $H(\mathbf{X}_m, t)$ is continuously differentiable, one can write

$$\begin{aligned} g_{l\,i,j}(\mathbf{Y}, \mathbf{X}, t) = & \sum_{l'=1}^{m}\sum_{i'=1}^{M}\sum_{j'=1}^{N} \frac{\partial h_{l\,i,j}(\mathbf{X}_m, t)}{\partial x_{l'\,i',j'}} \\ & \times f_{l'\,i',j'}(\mathbf{X}, \mathbf{Y}, t) + \frac{\partial h_{l\,i,j}(\mathbf{X}_m, t)}{\partial t} \\ & - q_{l\,i,j}(\mathbf{X}, \mathbf{Y}, t). \end{aligned}$$

Let

$$\mathbf{e} = H(\mathbf{X}_m, t) - \mathbf{Y}_m.$$

Then, the error equation has the following form:

$$\begin{aligned}
\frac{d\mathbf{e}}{dt} &= \sum\sum\left(\frac{dh_{l\,i,j}(\mathbf{X}_m,t)}{dt}\right)_{m\times M\times N} - \frac{d\mathbf{Y}_m}{dt} \\
&= \left(\sum_{l'=1}^{m}\sum_{i'=1}^{M}\sum_{j'=1}^{N}\frac{\partial h_{l\,i,j}(\mathbf{X}_m,t)}{\partial x_{l'\,i',j'}}\dot{x}_{l'\,i',j'}\right)_{m\times M\times N} \\
&\quad + \left(\frac{\partial h_{l\,i,j}(\mathbf{X}_m,t)}{\partial t}\right)_{m\times M\times N} \\
&\quad - (g_{l\,i,j}(\mathbf{Y},\mathbf{X},t))_{m\times M\times N} \\
&= \left(\sum_{l'=1}^{m}\sum_{i'=1}^{M}\sum_{j'=1}^{N}\frac{\partial h_{l\,i,j}(\mathbf{X}_m,t)}{\partial x_{l'\,i',j'}}\right. \\
&\quad \left.\times f_{l'\,i',j'}(\mathbf{X},\mathbf{Y},t)\right)_{m\times M\times N} + \left(\frac{\partial h_{l\,i,j}(\mathbf{X}_m,t)}{\partial t}\right)_{m\times M\times N} \\
&\quad - (g_{l\,i,j}(\mathbf{Y},\mathbf{X},t))_{m\times M\times N} \\
&= (q_{l\,i,j}(\mathbf{X},\mathbf{Y},t))_{m\times M\times N}\,.
\end{aligned}$$

By assumption, the two CCADS (1.2.86) and (1.2.87) are in PGC with respect to the transformations H. Therefore, $q_{l\,i,j}(\mathbf{Y},\mathbf{X},t)$ makes the zero solution of the error equation (2.3.32) be stable for any initial condition in the set B defined as in Definition 2.10.

$\Leftarrow$ If the function $g_{l\,i,j}(\mathbf{Y},\mathbf{X},t)$ can be represented by equation (2.3.31), and $(q_{l\,i,j}(\mathbf{X},\mathbf{Y},t))_{m\times M\times N}$ makes the zero solution of equation (2.3.32) be stable on the set B defined by Definition 2.10, then condition (2.2.10) holds. This means that the two CCADS (1.2.86) and (1.2.87) are in PGC with respect to the transformation $H(\mathbf{X}_m,t) = \mathbf{Y}_m$.

This completes the proof. □

Remark 2.3.3. In fact, $Q(\mathbf{Y}(t),\mathbf{X}(t),t) = (q_{l\,i,j}(\mathbf{X},\mathbf{Y},t))_{m\times M\times N}$ can be easily constructed. For example, take $Q(\mathbf{Y}(t),\ \mathbf{X}(t),t) = A\mathbf{e}(t)$, where A is a stable matrix; that is, its eigenvalues λ's have the following properties:

(1) $Re(\lambda) < 0$, or

(2) $Re(\lambda) = 0$ and it is simple.

2.4 Application of GC Theorems

2.4.1 *Application of GC Theorem to CDADS*

This subsection shows an application of Theorem 2.1 to the generalized consensus of CDADS. Based on the discrete Lorenz equation [Sprot (2003)], a coupled Lorenz CNN is first constructed.

Introduce a differentiable and invertible transformation H as follows:

$$H = \boldsymbol{B} \circ \tilde{H} : \mathbb{R}^{3\times 25\times 25} \to \mathbb{R}^{3\times 25\times 25} \tag{2.4.1}$$

where

$$\tilde{H} = (\tilde{h}_1, \tilde{h}_2, \tilde{h}_3) : \mathbb{R}^{3\times 25\times 25} \to \mathbb{R}^{3\times 25\times 25}, \tag{2.4.2}$$

and

$$\boldsymbol{B} = (\beta_{k,l})_{3\times 3} : \mathbb{R}^3 \to \mathbb{R}^3, \tag{2.4.3}$$

where

$$\begin{aligned}\tilde{h}_1((x_{1\,i,j})_{21\times 21}) &= (\alpha^1_{i,j})_{25\times 25}(x_{1\,i,j})_{25\times 25}\\ &= (\tilde{x}_{1\,i,j})_{25\times 25},\end{aligned} \tag{2.4.4}$$

$$\begin{aligned}\tilde{h}_2((x_{2\,i,j})_{25\times 25}) &= (\alpha^2_{i,j})_{25\times 25}(x_{2\,i,j})_{25\times 25}\\ &= (\tilde{x}_{2\,i,j})_{25\times 25},\end{aligned} \tag{2.4.5}$$

$$\begin{aligned}\tilde{h}_3((x_{3\,i,j})_{25\times 25}) &= (\alpha^2_{i,j})_{25\times 25}(x_{3\,i,j})_{25\times 25}\\ &= (\tilde{x}_{3\,i,j})_{25\times 25},\end{aligned} \tag{2.4.6}$$

and, for any triple $(\tilde{x}_{1\,i,j}, \tilde{x}_{2\,i,j}, \tilde{x}_{3\,i,j})$,

$$\boldsymbol{B}(\tilde{x}_{1\,i,j}, \tilde{x}_{2\,i,j}, \tilde{x}_{3\,i,j}) = (\beta_{k,l})_{3\times 3}[\tilde{x}_{1\,i,j}, \tilde{x}_{2\,i,j}, \tilde{x}_{3\,i,j}]^{\mathrm{T}} \tag{2.4.7}$$

where

$$\begin{cases}\boldsymbol{A}_1 = (\alpha^1_{i,j})_{25\times 25}\\ \boldsymbol{A}_2 = (\alpha^2_{i,j})_{25\times 25}\\ \boldsymbol{A}_3 = (\alpha^3_{i,j})_{25\times 25}\end{cases} \tag{2.4.8}$$

and

$$\boldsymbol{B} = \begin{bmatrix}1 & -1 & 1\\ 1 & 1 & 1\\ 0 & 1 & 1\end{bmatrix} \tag{2.4.9}$$

are all invertible matrices (see Fig. 2.4.1).

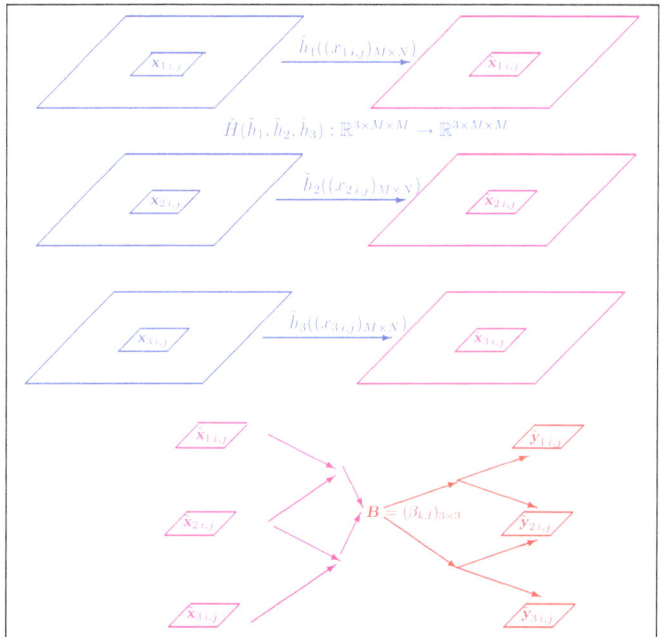

Fig. 2.4.1: Transformation $H = \boldsymbol{B} \circ \tilde{H} : \mathbb{R}^{3\times 25\times 25} \to \mathbb{R}^{3\times 25\times 25}$.

Based on Theorem 2.1, one can construct a GC system of Lorenz CNN as follows:

$$\begin{cases} x_{1\,i,j}(k+1) = x_{1\,i,j}(k)x_{2\,i,j}(k) - x_{3\,i,j}(k) \\ x_{2\,i,j}(k+1) = x_{1\,i,j}(k) \\ x_{3\,i,j}(k+1) = x_{2\,i,j}(k) - 0.001[0.2x_{3i+1,j} + 0.2x_{3i-1,j} \\ \qquad\qquad\qquad +0.2x_{3i,j+1} + 0.2x_{3i,j-1} - x_{3i,j}], \end{cases} \tag{2.4.10}$$

$$i, j = 1, 2, \cdots 25,$$

In a compact form, it can be rewritten as

$$\boldsymbol{X}(k+1) = F(\boldsymbol{X}(k)). \tag{2.4.11}$$

The second part of the Lorenz CNN has the following form:

$$\boldsymbol{Y}(k+1) = H(F(\boldsymbol{X}(k)) - Q(\boldsymbol{X}, \boldsymbol{Y}), \tag{2.4.12}$$

where H is defined by (2.4.1)–(2.4.9), and

$$Q(\boldsymbol{X}, \boldsymbol{Y}) = (q_{l\,i,j}(\boldsymbol{X}, \boldsymbol{Y}))_{3\times 25\times 25} \tag{2.4.13}$$

such that

$$(q_{l\,i,j}(\boldsymbol{X}, \boldsymbol{Y}))_{25\times 25} = C(e_{l\,i,j}(\boldsymbol{X}, \boldsymbol{Y}))_{25\times 25}, \tag{2.4.14}$$

$$l = 1, 2, 3,$$

in which

$$C = (c_{i,j})_{25\times 25} \tag{2.4.15}$$

$c_{1,1} = 1$ and, for other i, j :

$$c_{i,j} = \begin{cases} -0.5 & \text{if } i = j \\ 0 & \text{otherwise.} \end{cases} \tag{2.4.16}$$

By Theorem 2.1, systems (2.4.11) and (2.4.12) are in GC with respect to the transformation H defined by (2.4.1)–(2.4.9).

Now, select the following initial conditions:

$$(x_{l\,i,j}(0))_{25\times 25} = \mathbf{X}_0(l) + 0.02(rand(25, 25) - 0.5)$$
$$l = 1, 2, 3,$$

where $\mathbf{X}_0 = [0.5\ 0.5\ -1]^{\mathrm{T}}$, and $rand(25, 25)$ is a Matlab command which returns an 25×25 matrix containing pseudo-random values drawn from a uniform distribution on the unit interval.

$$\mathbf{Y}(0) = \mathbf{X}_0 + 0.02(rand(25, 25) - 0.5). \tag{2.4.17}$$

The chaotic trajectories of the components $x_{l\,13,12}$ and $y_{l\,13,12}$, $x_{l\,13,13}$ and $y_{l\,13,13}$, as well as $x_{l\,13,14}$ and $y_{l\,13,14}$, of the state variables $\mathbf{X}$ and $\mathbf{Y}$ of the first 2,000 iterations, are shown in Figs. 2.4.2–2.4.4, which shows clearly chaotic behaviors.

Figures 2.4.5(a)–(c) show that, although there are initial perturbations (2.4.17), the state variables $\mathbf{X}_{l,13,13}$ and $\mathbf{Y}_{l,13,13}$ achieve GC rapidly.

The chaotic orbits of some components $x_{l\,i,j}$, $y_{l\,i,j}$, and of the state variables $\mathbf{X}$ and $\mathbf{Y}$, are shown in Figs. 2.4.6(a)–(f). It can be observed that the dynamic behaviors of the neighboring cells at the lattice: (13, 12), (13, 13), and (13, 14), are quite different.

The three-dimensional views of the evolution of the discrete Lorenz CNN at different times are shown in Fig. 2.4.7, from which chaotic waves can been clearly seen. It can also be observed that the irregular chaotic waves shown by the first three columns in Fig. 2.4.7 have been firstly transformed to wall-shaped chaotic waveforms and finally become random waveforms as shown by the last three columns in Fig. 2.4.7.

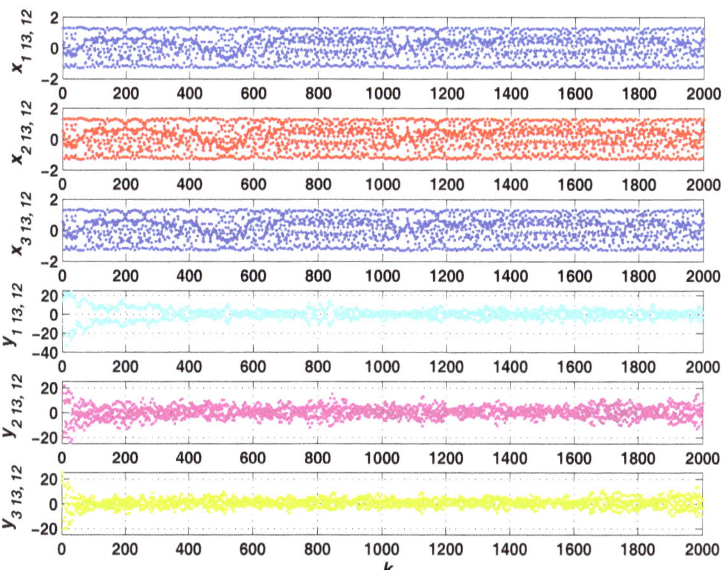

Fig. 2.4.2: Evolution of state variables: $t - x_{1\,13,12}$, $t - x_{2\,13,12}$, $t - x_{3\,13,12}$, $t - y_{1\,13,12}$, $t - y_{2\,13,12}$, and $t - y_{3\,1312}$.

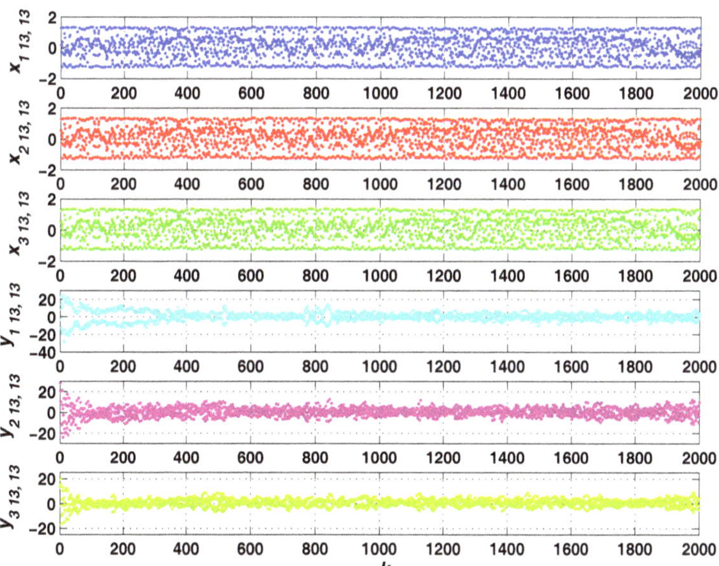

Fig. 2.4.3: Evolution of state variables: $t - x_{1\,13,13}$, $t - x_{2\,13,13}$, $t - x_{3\,13,13}$, $t - y_{1\,13,13}$, $t - y_{2\,13,13}$, and $t - y_{3\,1313}$.

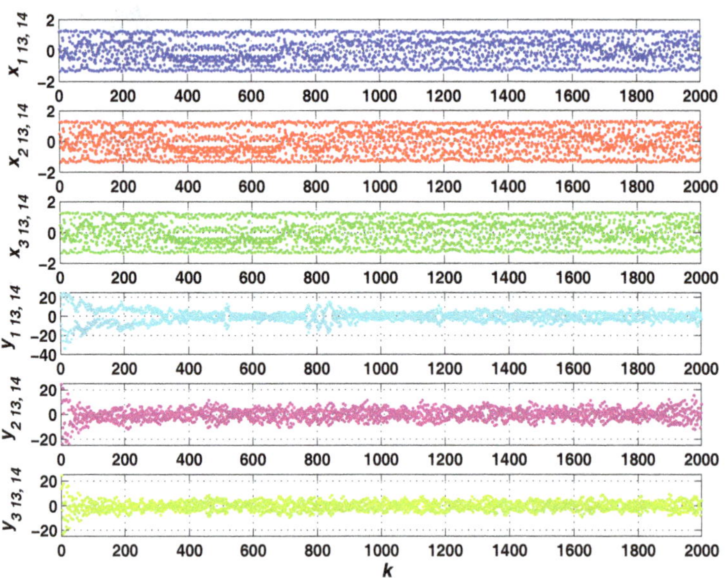

Fig. 2.4.4: Evolution of state variables: $t - x_{1\,13,14}$, $t - x_{2\,13,14}$, $t - x_{3\,13,14}$, $t - y_{1\,13,14}$, $t - y_{2\,13,14}$, and $t - y_{3\,1314}$.

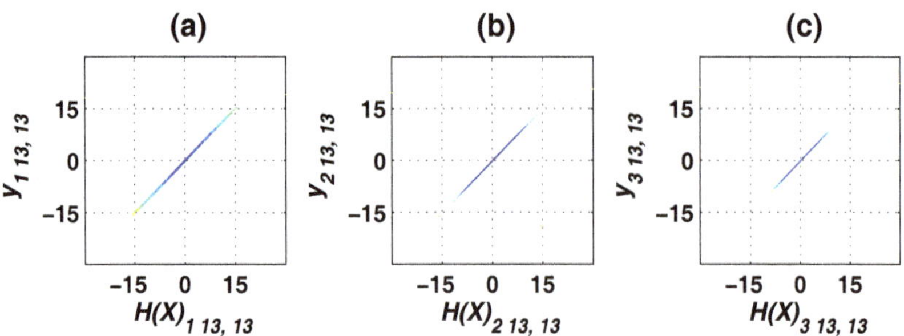

Fig. 2.4.5: (a), (b) and (c) show that the variables $x_{1\,13,13}$, $x_{2\,13,13}$, $x_{3\,13,13}$, and $y_{1\,13,13}$, $y_{2\,13,13}$, $y_{3\,13,13}$ are in chaos GC with respect to a transformation H.

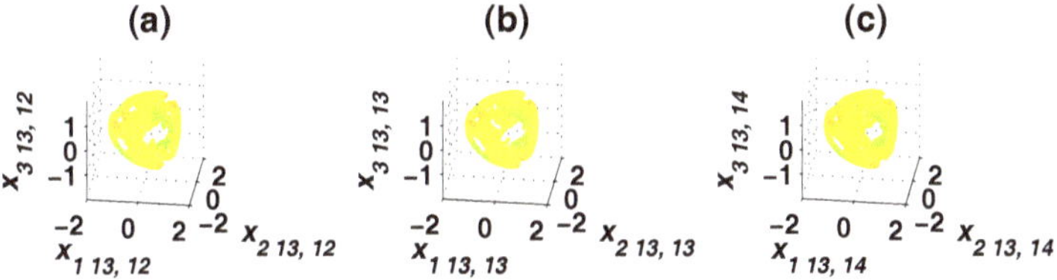

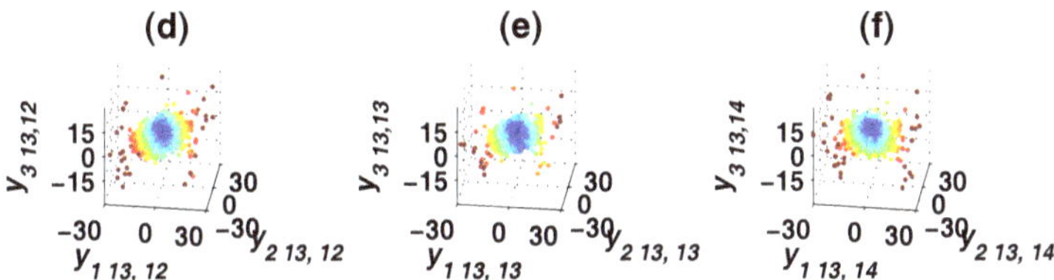

Fig. 2.4.6: Chaotic trajectories of some components of the state variables: (a) $x_{1\,13,12} - x_{2\,13,12} - x_{3\,13,12}$, (b) $x_{1\,13,13} - x_{2\,13,13} - x_{3\,13,13}$, (c) $x_{1\,13,14} - x_{2\,13,14} - x_{3\,13,14}$, (d) $y_{1\,13,12} - y_{2\,13,12} - y_{3\,13,12}$, (e) $y_{1\,13,13} - y_{2\,13,13} - y_{3\,13,13}$, (f) $y_{1\,13,14} - y_{2\,13,14} - y_{3\,13,14}$.

2.4.2 *Application of GC Theorem to Non-autonomous Bidirectional CDADS*

This subsection shows an application of Theorem 2.4 to the generalized consensus of CDADS.

A coupled discrete non-autonomous bidirectional CNN is constructed based on a perturbed 3D discrete Lorenz system. The first part of the GC system has the following form:

$$\begin{cases} x_{1\,i,j}(k+1) = x_{1\,i,j}(k)x_{2\,i,j}(k) - x_{3\,i,j}(k) \\ \qquad\qquad +0.001x_{1\,i,j}(k) + 5e-5\cos(\pi k) \\ x_{2\,i,j}(k+1) = x_{1\,i,j}(k) + 0.001x_{2\,i,j}(k) \\ \qquad\qquad +5e-5\cos(y_{1\,i,j}(k)y_{2\,i,j}(k)y_{3\,i,j}(k)) \\ x_{3\,i,j}(k+1) = x_{2\,i,j}(k) + 0.001x_{3\,i,j}(k) \\ \qquad\qquad -0.001[0.2x_{3i+1,j} + 0.2x_{3i-1,j} \\ \qquad\qquad +0.2x_{3i,j+1} + 0.2x_{3i,j-1} - x_{3i,j}], \end{cases} \quad (2.4.18)$$

$$i, j = 1, 2, \cdots, 25.$$

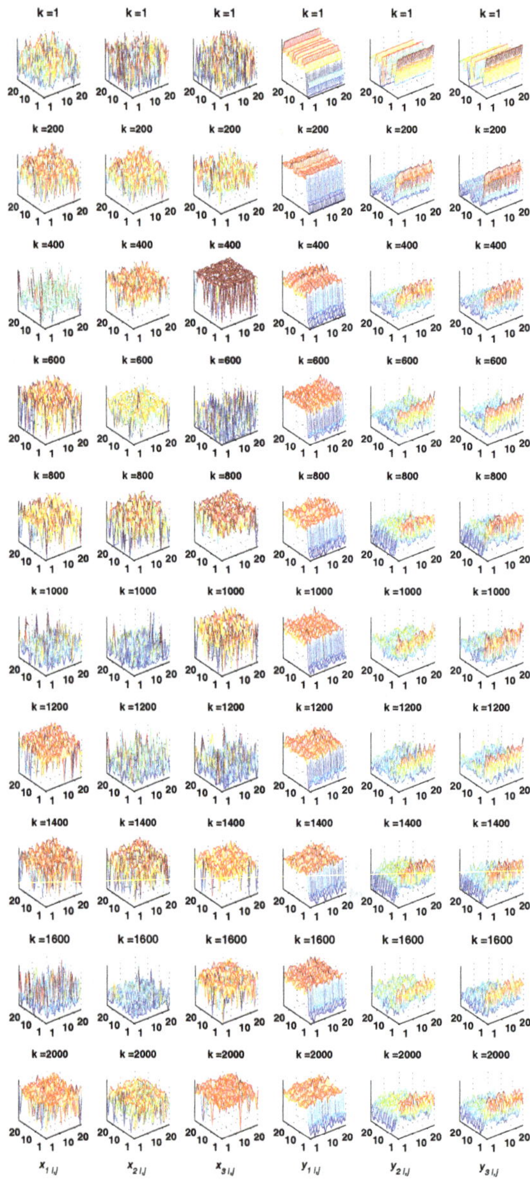

Fig. 2.4.7: Three-dimensional views of the Lorenz CNN spiral waves at different time t. The vertical axes represent the state variables $x_{1\,i,j}$, $x_{2\,i,j}$, $x_{3\,i,j}$, $y_{1\,i,j}$, $y_{2\,i,j}$, and $y_{3\,i,j}$, while the horizontal axes are the plane coordinates (i, j).

Firstly, construct a continuously differentiable transformation H as follows:

$$H = T \circ \boldsymbol{B} \circ \tilde{H} : \mathbb{R}^{3\times 25\times 25} \times \mathbb{Z}^+ \to \mathbb{R}^{3\times 25\times 25}, \tag{2.4.19}$$

where $\boldsymbol{B} \circ \tilde{H}$ is the same as that defined by (2.4.2) and (2.4.3). For any pair of $(\boldsymbol{X}, k) \in \mathbb{R}^{3\times 25\times 25} \times \mathbb{Z}^+$,

$$T(\boldsymbol{X}, k) = \boldsymbol{X} + 0.00005\cos(\pi k). \tag{2.4.20}$$

By Theorem 2.4, one can construct a perturbed Lorenz CNN with GC in a compact form: system (2.4.18) becomes

$$\boldsymbol{X}(k+1) = F(\boldsymbol{X}(k), \boldsymbol{Y}(k), k). \tag{2.4.21}$$

The second part of the perturbed Lorenz CNN has the following form:

$$\boldsymbol{Y}(k+1) = H(F(\boldsymbol{X}(k), \boldsymbol{Y}(k), k), k+1) - Q(\boldsymbol{X}(k), \boldsymbol{Y}(k), k) \tag{2.4.22}$$

where H is defined by (2.4.1)–(2.4.9), (2.4.19) and (2.4.20), and

$$Q(\boldsymbol{X}, \boldsymbol{Y}, k) = (q_{l\,i,j}(\boldsymbol{Y}, \boldsymbol{X}, k))_{3\times 25\times 25}, \tag{2.4.23}$$

such that

$$(q_{l\,i,j}(\boldsymbol{X}, \boldsymbol{Y}, k))_{l\times 25\times 25} = C(e_{l\,i,j}(\boldsymbol{X}, \boldsymbol{Y}))_{l\times 25\times 25}, \; l = 1, 2, 3, \tag{2.4.24}$$

in which

$$C = (c_{i,j})_{25\times 25} \tag{2.4.25}$$

$$C(1:3, 1:3) = \begin{bmatrix} 0 & 1 & 0 \\ 0 & 0 & 1 \\ 1 & -1 & 1 \end{bmatrix} \tag{2.4.26}$$

and, for other i, j,

$$c_{i,j} = \begin{cases} -0.5 & \text{if } i = j \\ 0 & \text{otherwise.} \end{cases} \tag{2.4.27}$$

By Theorem 2.4, systems (2.4.21) and (2.4.22) are in GC with respect to the transformation H.

Now, select the following initial conditions:

$$(x_{l\,i,j}(0))_{25\times 25} = \boldsymbol{X}_0(l) + 0.02(rand(25, 25) - 0.5), \quad l = 1, 2, 3.$$

where $\boldsymbol{X}_0 = [0.023471 \;\; -0.4371 \;\; 0.11459]^{\mathrm{T}}$.

$$\boldsymbol{Y}(0) = \boldsymbol{X}_0 + 0.02(rand(25, 25) - 0.5). \tag{2.4.28}$$

The chaotic trajectories of the components $x_{l\,13,12}$ and $y_{l\,13,12}$, $x_{l\,13,13}$ and $y_{l\,13,13}$, as well as $x_{l\,13,14}$ and $y_{l\,13,14}$ of the state variables $\boldsymbol{X}$ and $\boldsymbol{Y}$

of the first 2,000 iterations, are displayed in Figs 2.4.8–2.4.16, showing significant chaotic behaviors.

The chaotic orbits of some components $x_{l\,i,j}, y_{l\,i,j}$ and of state variables $\mathbf{X}$ and $\mathbf{Y}$ are shown in Figs. 2.4.11(a)–(f). It can be observed that the dynamic behaviors of the neighboring cells at the lattice: (13, 12), (13, 13), and (13, 14), are different.

Figures 2.4.12(a)–(c) show that, although there are initial perturbations (2.4.28), the state variables $\boldsymbol{X}_{l,13,13}$ and $\boldsymbol{Y}_{l,13,13}$ achieve GC rapidly.

The three-dimensional views of the evolution of the discrete perturbed Lorenz CNN at different times are shown in Fig. 2.4.13, where chaotic waves can been seen clearly. It can also be observed that the irregular chaotic waves shown by the first three columns in Fig. 2.4.13 have been firstly transformed to wall-shaped chaotic waveforms and finally become random waveforms as shown by the last three columns in Fig. 2.4.13.

2.4.3 *Application of PGC Theorem to Non-autonomous Bidirectional CDADS*

This subsection shows an application of Theorem 2.9 to non-autonomous bidirectional CDADS. Firstly, a non-autonomous bidirectional $8 \times 25 \times 25$ dimensional CNN with 8 state variables is introduced, which consists of simple sine functions.

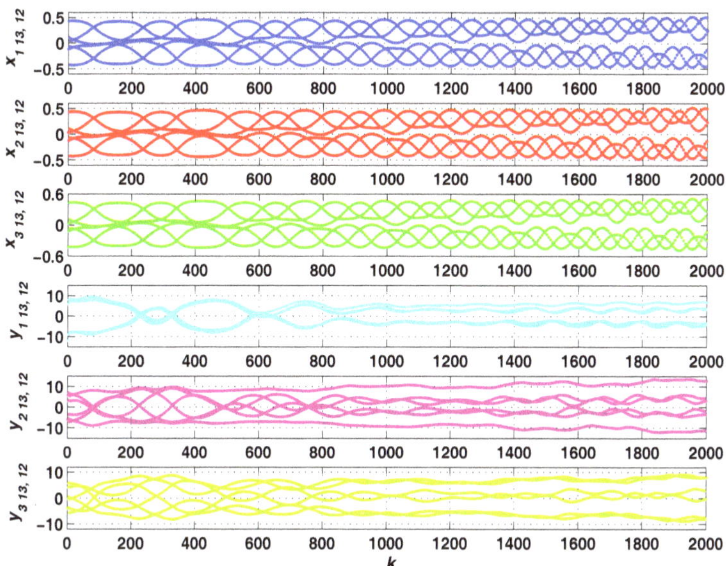

Fig. 2.4.8: Evolution of state variables: $t - x_{1\,13,12}$, $t - x_{2\,13,12}$, $t - x_{3\,13,12}$, $t - y_{1\,13,12}$, $t - y_{2\,13,12}$, and $t - y_{3\,1312}$.

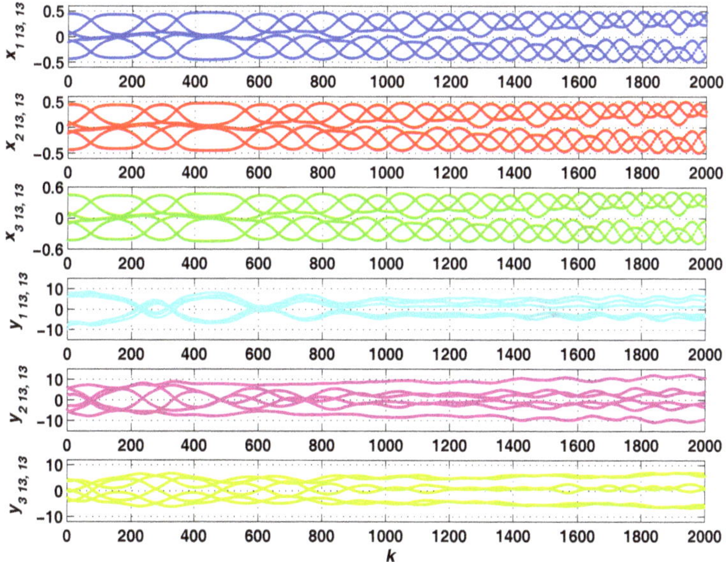

Fig. 2.4.9: Evolution of state variables: $t - x_{1\,13,13}$, $t - x_{2\,13,13}$, $t - x_{3\,13,13}$, $t - y_{1\,13,13}$, $t - y_{2\,13,13}$, and $t - y_{3\,1313}$.

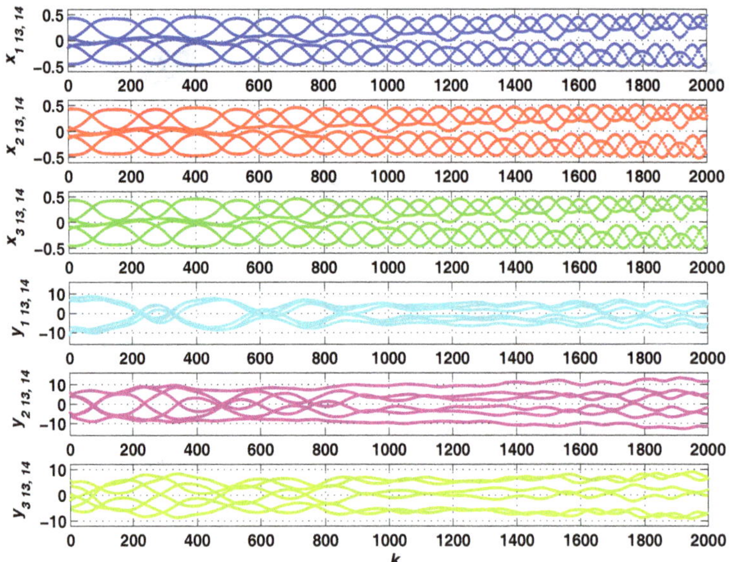

Fig. 2.4.10: Evolution of state variables: $t - x_{1\,13,14}$, $t - x_{2\,13,14}$, $t - x_{3\,13,14}$, $t - y_{1\,13,14}$, $t - y_{2\,13,14}$, and $t - y_{3\,1314}$.

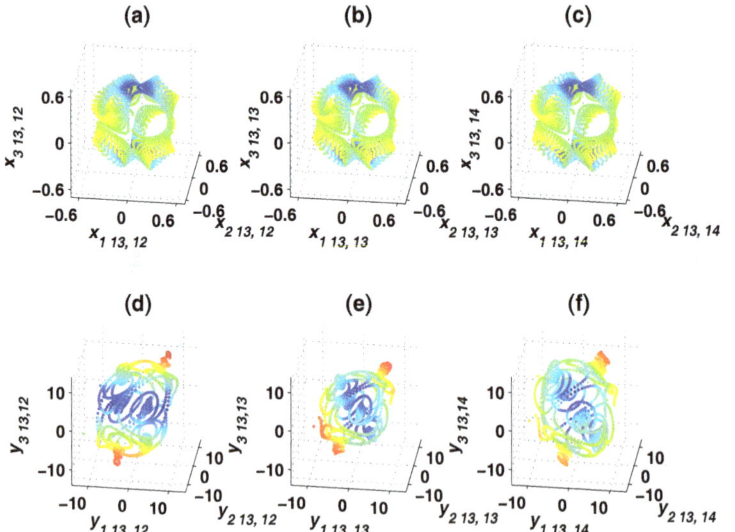

Fig. 2.4.11: Chaotic trajectories of some components of the state variables: (a) $x_{1\,13,12} - x_{2\,13,12} - x_{3\,13,12}$, (b) $x_{1\,13,13} - x_{2\,13,13} - x_{3\,13,13}$, (c) $x_{1\,13,14} - x_{2\,13,14} - x_{3\,13,14}$, (d) $y_{1\,13,12} - y_{2\,13,12} - y_{3\,13,12}$, (e) $y_{1\,13,13} - y_{2\,13,13} - y_{3\,13,13}$, (f) $y_{1\,13,14} - y_{2\,13,14} - y_{3\,13,14}$.

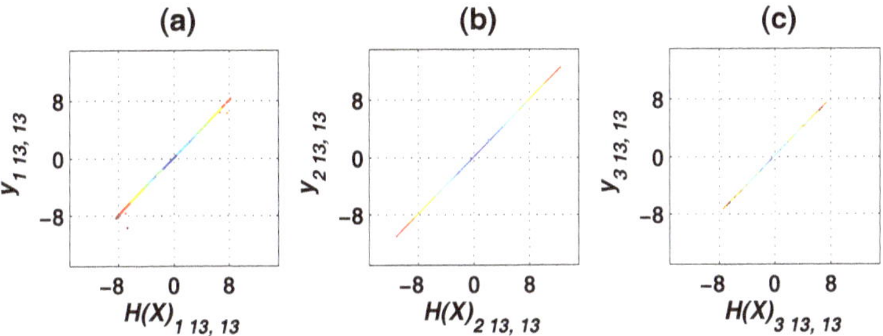

Fig. 2.4.12: (a), (b) and (c) show that the variables $x_{1\,13,13}$, $x_{2\,13,13}$, $x_{3\,13,13}$, and $y_{1\,13,13}$, $y_{2\,13,13}$, $y_{3\,13,13}$ are in chaos GC with respect to a transformation H.

The first part of the sine-function CNN has the following form:

$$\begin{cases} x_{1\,i,j}(k+1) = \sin(\pi x_{1\,i,j}(k)x_{2\,i,j}(k)) - \sin(\pi x_{3\,i,j}(k)) \\ x_{2\,i,j}(k+1) = \sin(\pi x_{1\,i,j}(k)) + \sin(\pi x_{4\,i,j}(k)) \\ x_{3\,i,j}(k+1) = \sin(\pi x_{2\,i,j}(k)) \\ x_{4\,i,j}(k+1) = \sin(\pi x_{2i,j}x_{3i,j}) + \sin(\pi k/2) \\ \qquad\qquad +0.001\sin(\pi y_{4i,j}) \\ \qquad\qquad -D[0.2x_{4i+1,j} + 0.2x_{4i-1,j} \\ \qquad\qquad +0.2x_{4i,j+1} + 0.2x_{4i,j-1} - x_{4i,j}], \end{cases} \tag{2.4.29}$$

$$i,j = 1,2,\cdots,25.$$

Now, construct a continuously differentiable transformation H as follows:

$$H = T \circ \boldsymbol{B} \circ \tilde{H} : \mathbb{R}^{3\times25\times25} \times \mathbb{Z}^+ \to \mathbb{R}^{3\times25\times25} \tag{2.4.30}$$

where

$$\tilde{H} = (\tilde{h}_1, \tilde{h}_2, \tilde{h}_3) : \mathbb{R}^{3\times21\times21} \to \mathbb{R}^{3\times25\times25}, \tag{2.4.31}$$

$$\boldsymbol{B} : \mathbb{R}^3 \to \mathbb{R}^3, \tag{2.4.32}$$

and

$$T : \mathbb{R}^{3\times25\times25} \times \mathbb{R}^+ \to \mathbb{R}^{3\times25\times25} \times \mathbb{R}^+ \tag{2.4.33}$$

such that

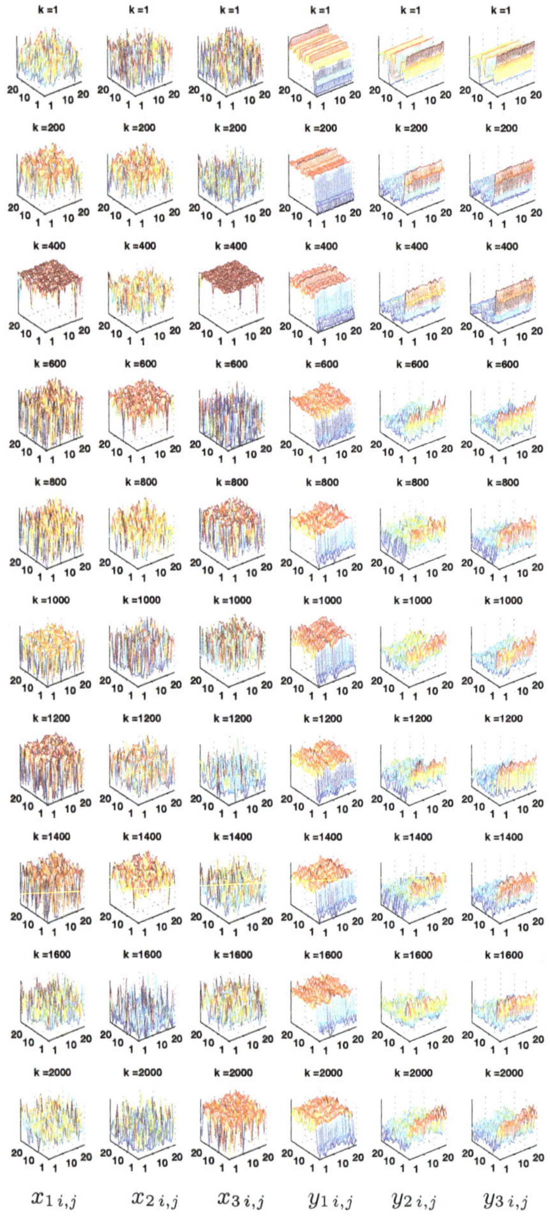

Fig. 2.4.13: The three-dimensional views of the non-autonomous bidirectional Lorenz CNN spiral waves at different time t. The vertical axes represent the state variables $x_{1\,i,j}$, $x_{2\,i,j}$, $x_{3\,i,j}$, $y_{1\,i,j}, y_{2\,i,j}$, and $y_{3\,i,j}$, while the horizontal axes are the plane coordinates (i, j).

$$\begin{aligned}\tilde{h}_1((x_{1\,i,j})_{25\times 25}) &= (\alpha^1_{i,j})_{25\times 25}(x_{1\,i,j})_{25\times 25}\\ &= (\tilde{x}_{1\,i,j})_{25\times 25},\end{aligned} \tag{2.4.34}$$

$$\begin{aligned}\tilde{h}_2((x_{2\,i,j})_{25\times 25}) &= (\alpha^2_{i,j})_{25\times 25}(x_{2\,i,j})_{25\times 25}\\ &= (\tilde{x}_{2\,i,j})_{25\times 25},\end{aligned} \tag{2.4.35}$$

$$\begin{aligned}\tilde{h}_3((x_{3\,i,j})_{25\times 25}) &= (\alpha^3_{i,j})_{25\times 25}(x_{3\,i,j})_{25\times 25}\\ &= (\tilde{x}_{3\,i,j})_{25\times 25},\end{aligned} \tag{2.4.36}$$

where

$$\begin{aligned}\boldsymbol{A}_1 &= (\alpha^1_{i,j})_{25\times 25}\\ &= \begin{cases} 0 & \text{if } i < j\\ 1/25 & \text{otherwise}\end{cases}\end{aligned} \tag{2.4.37}$$

$$\begin{aligned}\boldsymbol{A}_2 &= (\alpha^2_{i,j})_{25\times 25}\\ &= \begin{cases} 1/25 & \text{if } i = j\\ 1/100 & \text{if } j = i+1\\ 0 & \text{otherwise}\end{cases}\end{aligned} \tag{2.4.38}$$

$$\begin{aligned}\boldsymbol{A}_3 &= (\alpha^3_{i,j})_{25\times 25}\\ &= \begin{cases} 0 & \text{if } j < i\\ 1/25 & \text{otherwise}\end{cases}\end{aligned} \tag{2.4.39}$$

are invertible matrices. For any triple $(\tilde{x}_{1\,i,j}, \tilde{x}_{2\,i,j}, \tilde{x}_{2\,i,j})$, one has

$$\boldsymbol{B}(\tilde{x}_{1\,i,j}, \tilde{x}_{2\,i,j}, \tilde{x}_{3\,i,j}) = (\beta_{i,j})_{3\times 3}\tilde{\mathbf{X}} \tag{2.4.40}$$

$$= \begin{bmatrix} 1 & -1 & 1\\ 1 & 1 & 1\\ 0 & 1 & 1\end{bmatrix} \begin{bmatrix} (\tilde{x}_{1\,i,j})\\ (\tilde{x}_{2\,i,j})\\ (\tilde{x}_{3\,i,j})\end{bmatrix}, \tag{2.4.41}$$

and, for any pair $(\mathbf{X}, k) \in \mathbb{R}^{3\times 25\times 25} \times \mathbb{Z}^+$, one has

$$T(\mathbf{X}, k) = \mathbf{X} + 0.0005\sin(\pi k/2). \tag{2.4.42}$$

Based on Theorem 2.9, one can construct the sine-function CNN as follows: In a compact form, system (2.4.29) is written as

$$\mathbf{X}(k+1) = F(\mathbf{X}(k),\ \mathbf{Y}(k), k). \tag{2.4.43}$$

The second part of the sine-function CNN has the following form:

$$\begin{aligned}&\mathbf{Y}(k+1) = G(\mathbf{Y}(k),\ \mathbf{X}(k), k) =\\ &\begin{bmatrix} H(F_3(\mathbf{X}(k), \mathbf{Y}(k), k+1) - Q(\mathbf{X}(k), \mathbf{Y}(k), k)\\ (2\sin(\pi y_{4,i,j}))_{25\times 25}\end{bmatrix}\end{aligned} \tag{2.4.44}$$

where H is defined by (2.4.30)–(2.4.42), and

$$Q(\mathbf{X},\mathbf{Y},k) = (q_{l\,i,j}(\mathbf{X},\mathbf{Y},k))_{3\times 25\times 25} \tag{2.4.45}$$

such that

$$(q_{l\,i,j}(\mathbf{X},\mathbf{Y},k))_{25\times 25} = C(e_{l\,i,j}(k))_{25\times 25},\quad l=1,2,3, \tag{2.4.46}$$

where

$$C = (c_{i,j})_{25\times 25} \tag{2.4.47}$$

such that

$$C(1:3,1:3) = \begin{bmatrix} 0 & 1 & 0 \\ 0 & 0 & 1 \\ 1 & -1 & 1 \end{bmatrix}. \tag{2.4.48}$$

For other i, j,

$$c_{i,j} = \begin{cases} -0.5 & \text{if } i=j \\ 0 & \text{otherwise.} \end{cases} \tag{2.4.49}$$

It follows from Theorem 2.9 that systems (2.4.43) and (2.4.44) are in GC with respect to the transformation H.

Now, select the following initial conditions:

$$(x_{l\,i,j}(0))_{25\times 25} = \mathbf{X}_0(l) + 0.02(rand(25,25)-0.5), l=1,2,3,4,$$

where $\mathbf{X}_0 = [-0.79296\ \ 0.92414\ \ 0.6198\ \ -0.61394]^{\mathrm{T}}$.

$$\mathbf{Y}(0) = \mathbf{Y}_0 + 0.02(rand(25,25)-0.5), \tag{2.4.50}$$

where $\mathbf{Y}_0 = [0.54964\ \ 1.3526\ \ -0.37151\ \ 0.27054]^{\mathrm{T}}$.

The chaotic trajectories of the components $x_{l\,13,12}$ and $y_{l\,13,12}$, $x_{l\,13,13}$ and $y_{l\,13,13}$, as well as $x_{l\,13,14}$ and $y_{l\,13,14}$, of the state variables $\mathbf{X}$ and $\mathbf{Y}$ of the first 2,000 iterations, are shown in Figs. 2.4.14–2.4.16, which shows clear chaotic behaviors–the orbital points of the first state variables are distributed "homogeneously" in the hypercubs with both upper and lower bounds equal to 2.

Figures 2.4.17(a)–(c) show that, although there are initial perturbations (2.4.50), the state variables $\mathbf{X}_{l,13,13}$ and $\mathbf{Y}_{l,13,13}$ achieve GC rapidly.

The chaotic orbits of some components $x_{l\,i,j}$, $y_{l\,i,j}$ and of the state variables $\mathbf{X}$ and $\mathbf{Y}$ are shown in Figs. 2.4.18(a)–(f). It can be observed that the dynamic behaviors of the neighboring cells at the lattice: (13, 12), (13, 13), and (13, 14) are quite different. The transformation H changes the hypercube-shaped orbit domains of the state variable $\mathbf{X}$ to a narrower ellipsoid-shaped orbit domain of the state variable $\mathbf{Y}$.

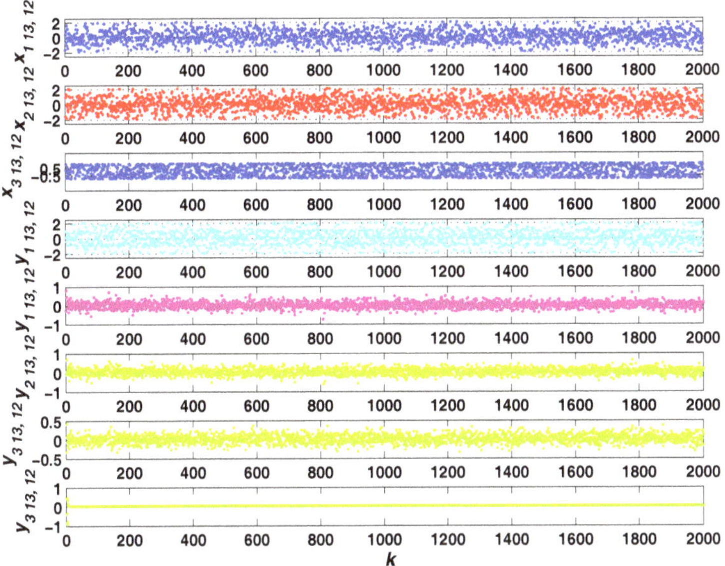

Fig. 2.4.14: Evolution of state variables: $t - x_{1\,13,12}$, $t - x_{2\,13,12}$, $t - x_{3\,13,12}$, $t - x_{4\,1312}$, $t - y_{1\,13,12}$, $t - y_{2\,13,12}$, $t - y_{3\,1312}$, and $t - y_{4\,1312}$.

The three-dimensional views of the evolution of the sine-function CNN at different times are shown in Fig. 2.4.19, in which chaotic waves can been seen clearly. It can be observed that the irregular chaotic waves shown by the first three columns in Fig. 2.4.19 have been firstly transformed to wall-shaped chaotic waveforms and finally become random waveforms as shown by the last three columns in Fig. 2.4.19.

2.4.4 *Application of GC Theorem to CCADS*

This subsection shows an application of Theorem 2.5 to CCADS.

Firstly, based on the third-order Lu-Chen chaotic system [Lü and Chen (2002)], an autonomous CNN is introduced as follows:

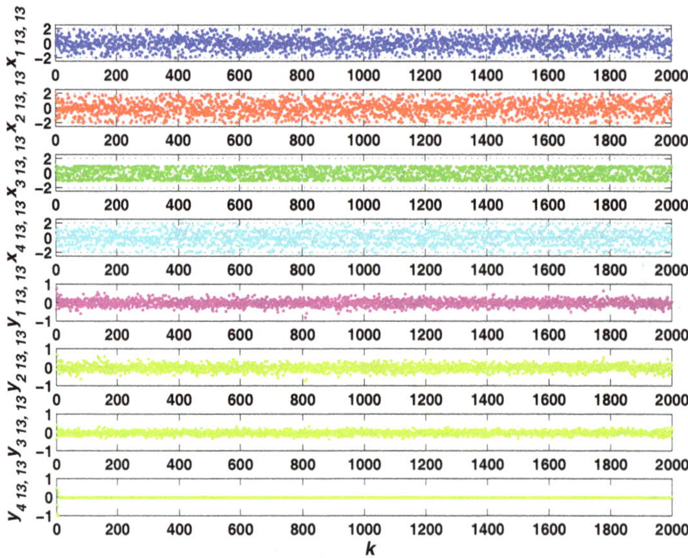

Fig. 2.4.15: Evolution of state variables: $t - x_{1\,13,13}$, $t - x_{2\,13,13}$, $t - x_{3\,13,13}$, $t - y_{1\,13,13}$, $t - y_{2\,13,13}$, and $t - y_{3\,1313}$.

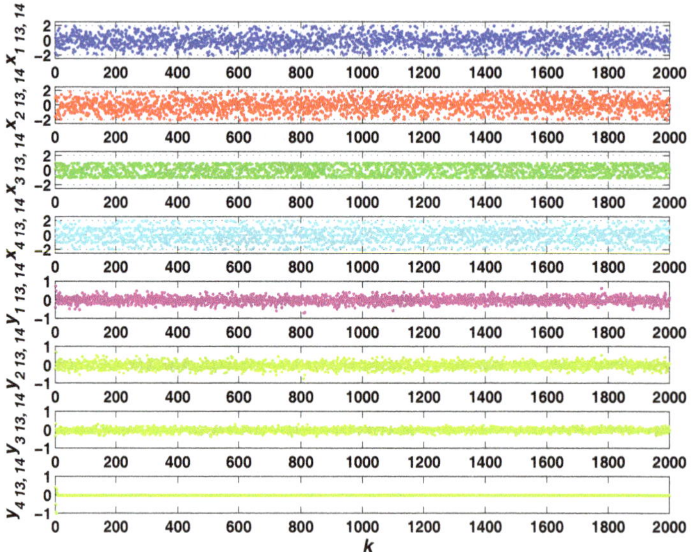

Fig. 2.4.16: Evolution of state variables: $t - x_{1\,13,14}$, $t - x_{2\,13,14}$, $t - x_{3\,13,14}$, $t - y_{1\,13,14}$, $t - y_{2\,13,14}$, and $t - y_{3\,1314}$.

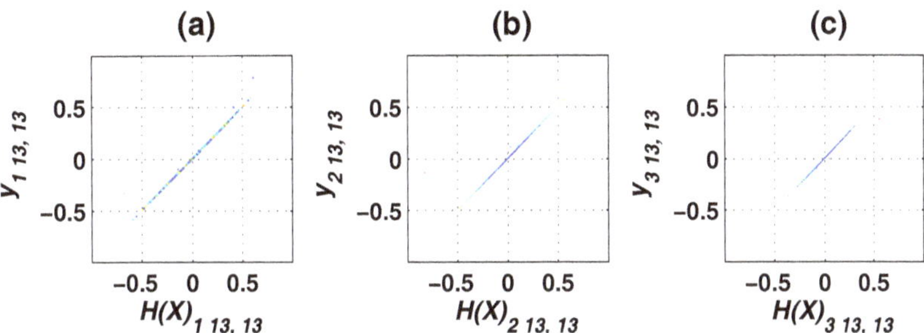

Fig. 2.4.17: (a), (b) and (c) show that the variables $x_{1\,13,13}$, $x_{2\,13,13}$, $x_{3\,13,13}$, and $y_{1\,13,13}$, $y_{2\,13,13}$, $y_{3\,13,13}$ are in chaos GC with respect to a transformation H.

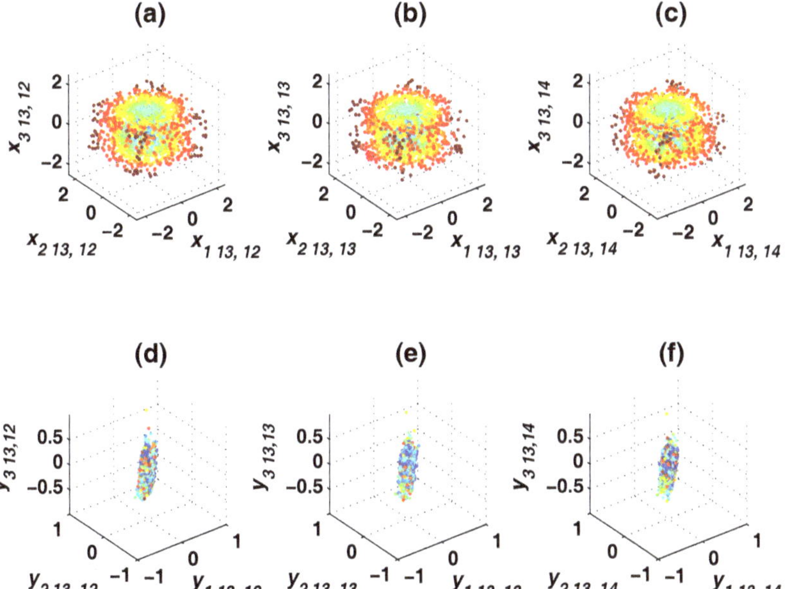

Fig. 2.4.18: Chaotic trajectories of some components of the state variables: (a) $x_{1\,13,12} - x_{2\,13,12} - x_{3\,13,12}$, (b) $x_{1\,13,13} - x_{2\,13,13} - x_{3\,13,13}$, (c) $x_{1\,13,14} - x_{2\,13,14} - x_{3\,13,14}$, (d) $y_{1\,13,12} - y_{2\,13,12} - y_{3\,13,12}$, (e) $y_{1\,13,13} - y_{2\,13,13} - y_{3\,13,13}$, (f) $y_{1\,13,14} - y_{2\,13,14} - y_{3\,13,14}$.

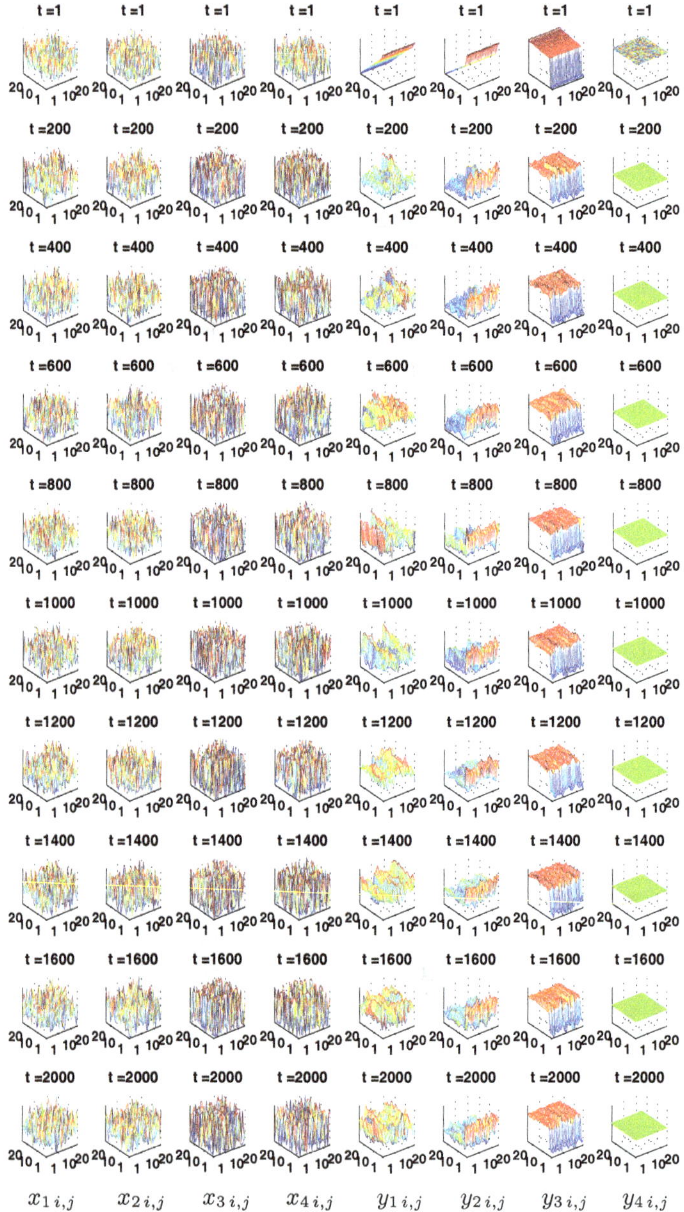

Fig. 2.4.19: The three-dimensional views of the non-autonomous bidirectional Lorenz CNN spiral waves at different time t. The vertical axes represent the state variables $x_{1\,i,j}$, $x_{2\,i,j}$, $x_{3\,i,j}$, $x_{4\,i,j}$, $y_{1\,i,j}$, $y_{2\,i,j}$, $y_{3\,i,j}$, and $y_{4\,i,j}$, while the horizontal axes are the plane coordinates (i, j).

$$\begin{cases} \dot{x}_{1i,j} = a(x_{2i,j} - x_{1i,j}) \\ \dot{x}_{2i,j} = -x_{1i,j}x_{3i,j} + cx_{2i,j} \\ \dot{x}_{3i,j} = x_{1i,j}x_{2i,j} - bx_{3i,j} + D[-0.2x_{3i+1,j} \\ \qquad -0.2x_{3i-1,j} - 0.2x_{3i,j+1} - 0.2x_{3i,j-1} + x_{3i,j}], \end{cases} \tag{2.4.51}$$

$$i, j = 1, 2, \ldots, 25,$$

where $a = 35, b = 3, c = 28, k_0 = 0.1, D_i = 0.0005$.

In a compact form, the CCADS (2.4.51) can be written as

$$\dot{\boldsymbol{X}} = F(\boldsymbol{X}, \boldsymbol{Y}). \tag{2.4.52}$$

Now, let

$$\dot{\boldsymbol{Y}} = G(\boldsymbol{Y}, \boldsymbol{X}), \tag{2.4.53}$$

where

$$g_{l\,i,j}(\boldsymbol{Y}, \boldsymbol{X}, t) = \sum_{l'=1}^{3} \sum_{i'=1}^{25} \sum_{j'=1}^{25} \frac{\partial h_{l\,i,j}(\boldsymbol{X})}{\partial x_{l'\,i',j'}} f_{l'\,i',j'}(\boldsymbol{X}, \boldsymbol{Y}) \\ -q_{l\,i,j}(\boldsymbol{X}, \boldsymbol{Y}), \tag{2.4.54}$$

$$H(\boldsymbol{X}) = (h_{l\,i,j}(\boldsymbol{X}))_{3\times 25\times 25} \\ = \left(\sum_{h=1}^{3} \beta_{l,h} \sum_{k=1}^{25} \sum_{m=1}^{25} \alpha_{h\,i,k} x_{h\,m,j} \right)_{3\times 25\times 25}. \tag{2.4.55}$$

Here,

$$\boldsymbol{A}_l = (\alpha_{l\,i,k})_{\times 25\times 25}, \quad l = 1, 2, 3, \tag{2.4.56}$$

$$\boldsymbol{B} = (\beta_{i,k})_{3\times 3} \tag{2.4.57}$$

are invertible matrices, and

$$q_{l\,i,j}(\boldsymbol{X}, \boldsymbol{Y}) = \begin{cases} 0 & \text{if } l = i = j = 0, \\ h_{l\,i,j}(\boldsymbol{X}) - y_{l\,i,j} & \text{otherwise.} \end{cases} \tag{2.4.58}$$

It follows from Theorem 2.5 that systems (2.4.52) and (2.4.53) achieve GC with respect to the transformation H defined by (2.4.55)–(2.4.57).

Next, select the following initial conditions:

$$(x_{l\,i,j}(0))_{25\times 25} = \boldsymbol{X}_0(l) + 0.02(rand(25, 25) - 0.5), \\ l = 1, 2, 3,$$

$$\boldsymbol{Y}(0) = \boldsymbol{X}(0) + 0.02(rand(21, 21) - 0.5), \tag{2.4.59}$$

where $\boldsymbol{X}_0 = [10.169\ 11.614\ 18.252]^{\mathrm{T}}$.

The chaotic trajectories of the components $x_{k\,13,12}$ and $y_{k\,13,12}$, $x_{k\,13,13}$ and $y_{k\,13,13}$, as well as $x_{k\,13,14}$ and $y_{k\,13,14}$, of the state variables $\boldsymbol{X}$ and $\boldsymbol{Y}$ over the time interval [0, 20], are shown in Figs. 2.4.20–2.4.22, which displays clear chaotic behaviors.

Figures 2.4.23(a)–(c) show the position relationships for $\boldsymbol{X}_{l,13,13}$ and $\boldsymbol{Y}_{l,13,13}$ by the transformation H.

Figures 2.4.23(d)–(f) show the ones after $t = 12.612$, which means that the trajectories of the components of variables $\boldsymbol{X}$ and $\boldsymbol{Y}$ enter the basin of attraction of a stable point.

The chaotic orbits of some components $x_{l\,i,j}$, $y_{l\,i,j}$ and of the state variables **X** and **Y** are shown in Fig. 2.4.24. It can be observed that the dynamic behaviors of the neighboring cells at the lattices (13, 12), (13, 13), and (13, 14), are quite different.

The three-dimensional views of the evolution of the Lu-Chen chaotic attractor CNN at different times are shown in Fig. 2.4.25, in which chaotic waves can been seen clearly. It can be observed that the irregular chaotic waves shown by the first three columns in Fig. 2.4.25 have been transformed to wall-shaped chaotic waveforms shown by the last three columns in Fig. 2.4.25.

2.4.5 *Application of GC Theorem to Non-autonomous Bidirectional CCADS*

This subsection shows an application of Theorem 2.9 to non-autonomous bidirectional CCADS.

First, based on the third-order Lorenz equation [Sparrow (1982)], a non-autonomous bidirectional Lorenz CNN is introduced as follows:

$$\begin{cases} \dot{x}_{1i,j} = a(x_{2i,j} - x_{1i,j}) + k_0 \sin(\pi t) \\ \dot{x}_{2i,j} = cx_{1i,j} - x_{1i,j}x_{3i,j} + cx_{2i,j} + D\cos(\pi y_{1i,j}y_{2i,j}y_{3i,j}) \\ \dot{x}_{3i,j} = x_{1i,j}x_{2i,j} - bx_{3i,j} + D[0.2x_{3i+1,j} + 0.2x_{3i-1,j} \\ \qquad +0.2x_{3i,j+1} + 0.2x_{3i,j-1} - x_{3i,j}], \end{cases} \tag{2.4.60}$$

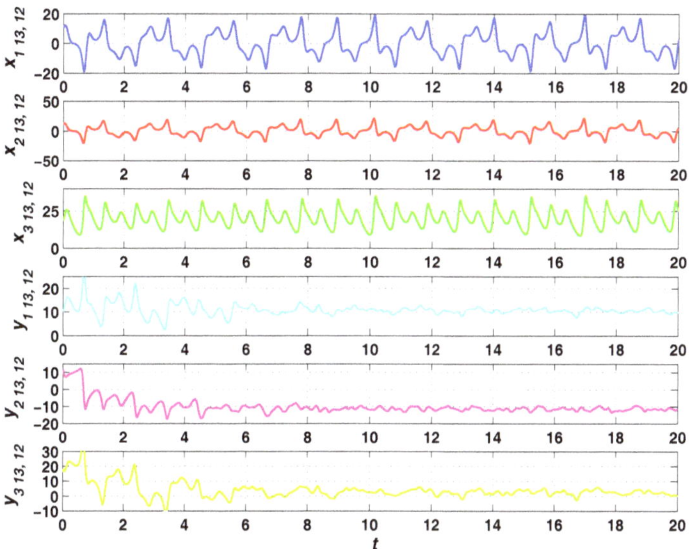

Fig. 2.4.20: Evolution of state variables: $t - x_{1\,13,12}$, $t - x_{2\,13,12}$, $t - x_{3\,13,12}$, $t - y_{1\,13,12}$, $t - y_{2\,13,12}$, and $t - y_{3\,1312}$.

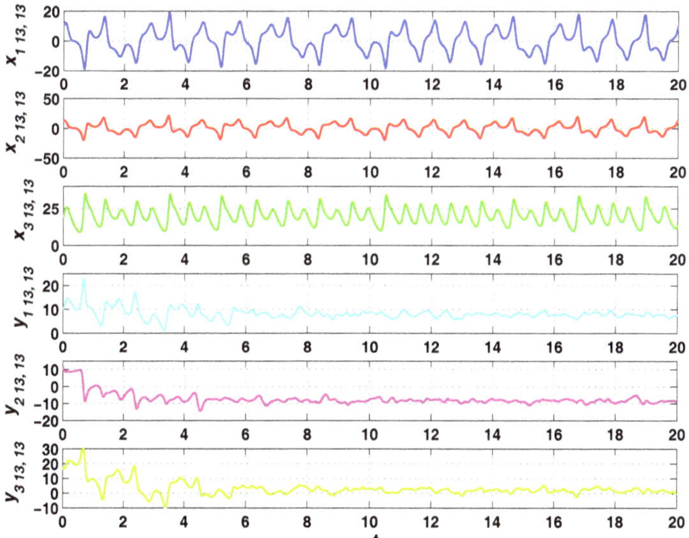

Fig. 2.4.21: Evolution of state variables: $t - x_{1\,13,13}$, $t - x_{2\,13,13}$, $t - x_{3\,13,13}$, $t - y_{1\,13,13}$, $t - y_{2\,13,13}$, and $t - y_{3\,1313}$.

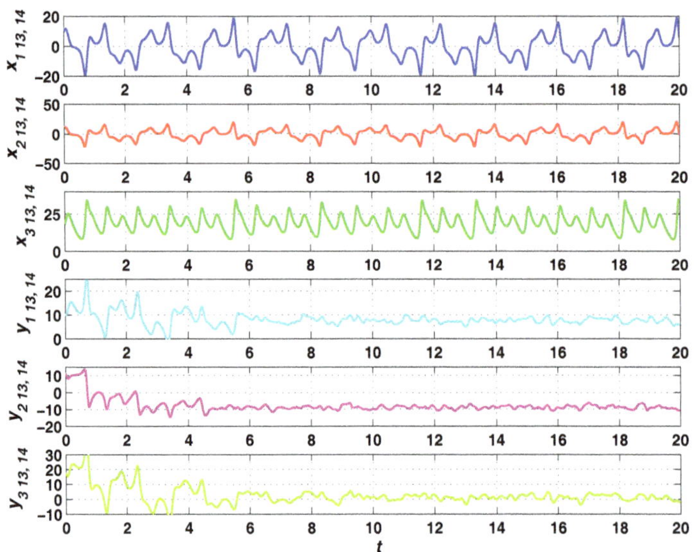

Fig. 2.4.22: Evolution of state variables: $t-x_{1\,13,14}$, $t-x_{2\,13,14}$, $t-x_{3\,13,14}$, $t-y_{1\,13,14}$, $t-y_{2\,13,14}$, and $t-y_{3\,1314}$.

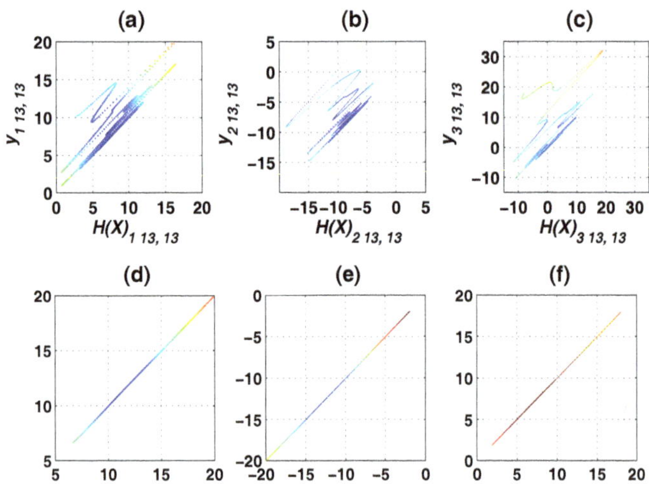

Fig. 2.4.23: (a), (b) and (c) show that the variables $x_{1\,13,13}$, $x_{2\,13,13}$, $x_{3\,13,13}$, and $y_{1\,13,13}$, $y_{2\,13,13}$, $y_{3\,13,13}$ are in chaos GC with respect to a transformation H defined by (2.4.55)–(2.4.57). (d), (e), (f) show that the corresponding relationships after $t = 12.612$.

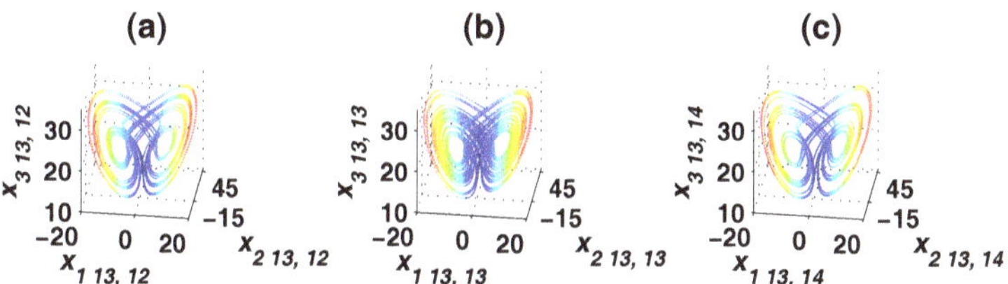

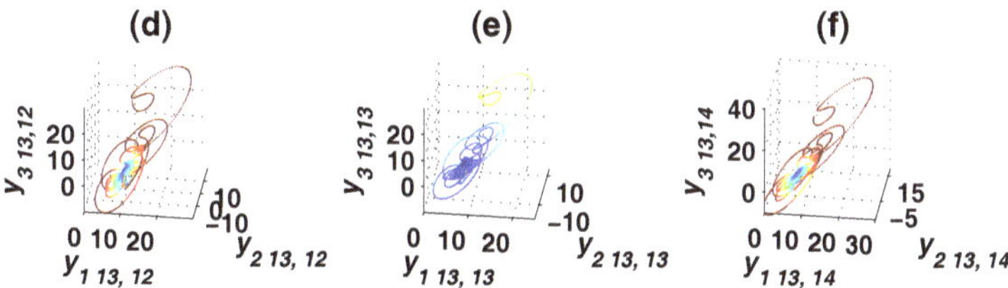

Fig. 2.4.24: Chaotic trajectories of some components of the state variables: (a) $x_{1\,13,12} - x_{2\,13,12} - x_{3\,13,12}$, (b) $x_{1\,13,13} - x_{2\,13,13} - x_{3\,13,13}$, (c) $x_{1\,13,14} - x_{2\,13,14} - x_{3\,13,14}$, (d) $y_{1\,13,12} - y_{2\,13,12} - y_{3\,13,12}$, (e) $y_{1\,13,13} - y_{2\,13,13} - y_{3\,13,13}$, (f) $y_{1\,13,14} - y_{2\,13,14} - y_{3\,13,14}$.

where $i, j = 1, 2, \ldots, 25, a = 35, b = 3, c = 28, k_0 = 0.1, D_i = 5 \times 10^{-4}$.

In a compact form, the CCADS (2.4.60) can be written as

$$\dot{\mathbf{X}} = F(\mathbf{X}, \mathbf{Y}, t). \tag{2.4.61}$$

Now, let

$$\dot{\mathbf{Y}} = G(\mathbf{Y}, \mathbf{X}, t), \tag{2.4.62}$$

where

$$\begin{aligned} g_{l\,i,j}(\mathbf{Y}, \mathbf{X}, t) = \sum_{l'=1}^{3} \sum_{i'=1}^{25} \sum_{j'=1}^{25} & \frac{\partial h_{l\,i,j}(\mathbf{X})}{\partial x_{l'\,i',j'}} [f_{l'\,i',j'}(\mathbf{X}, \mathbf{Y}, t) \\ & + 5e - 5\pi \cos(\pi t)] \\ & - q_{l\,i,j}(\mathbf{X}, \mathbf{Y}, t), \end{aligned} \tag{2.4.63}$$

$$\begin{aligned} H(\mathbf{X}, t) &= (h_{l\,i,j}(\mathbf{X}))_{3\times 25\times 25} + 5e - 5\sin(\pi t) \\ &= \left(\sum_{h=1}^{3} \beta_{l,h} \sum_{k=1}^{25} \sum_{m=1}^{25} \alpha_{h\,i,k} x_{h\,m,j} \right)_{3\times 25\times 25} + 5e - 5\sin(\pi t). \end{aligned} \tag{2.4.64}$$

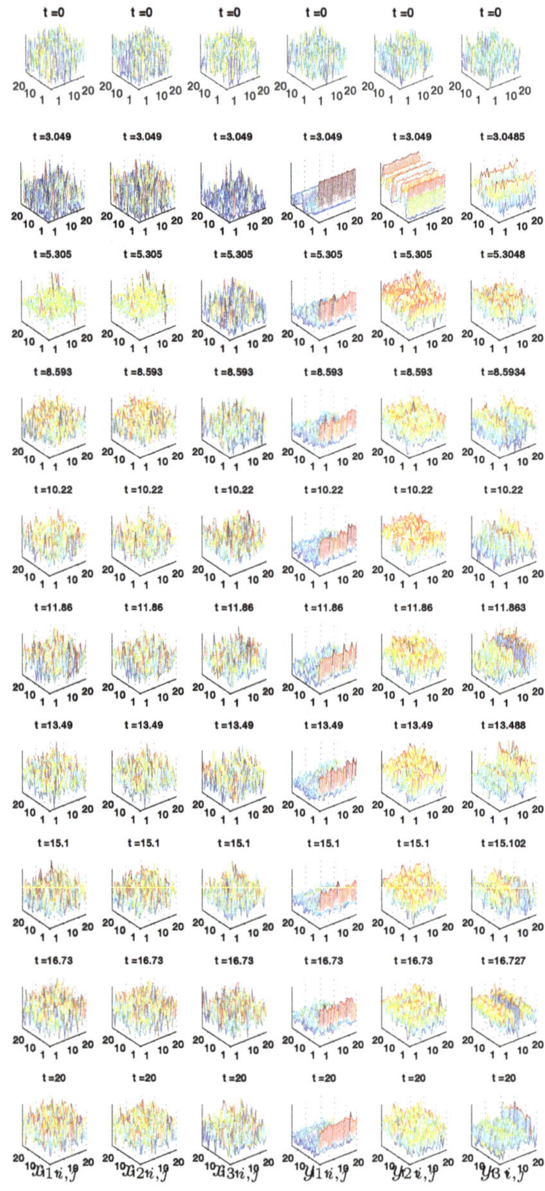

Fig. 2.4.25: The three-dimensional views of the non-autonomous bidirectional Chen CNN spiral waves at different time t. The vertical axes represent the state variables $x_{1\,i,j}$, $x_{2\,i,j}$, $x_{3\,i,j}$, $y_{1\,i,j}$, $y_{2\,i,j}$, $y_{3\,i,j}$, while the horizontal axes are the plane coordinates (i, j).

Here,

$$\boldsymbol{A}_l = (\alpha_{l\,i,k})_{\times 25\times 25}, \quad l = 1, 2, 3, \tag{2.4.65}$$

$$\boldsymbol{B} = (\beta_{l,h}) = \begin{bmatrix} 0.001 & -0.039 & 0.037 \\ 0.042 & -0.062 & -0.04 \\ -0.014 & -0.062 & 0.008 \end{bmatrix} \tag{2.4.66}$$

are invertible matrices, and

$$q_{l\,i,j}(\mathbf{X}, \mathbf{Y}) = \begin{cases} 1e - 5\sin(\pi t) \text{ if } l, i, j = 0, \text{otherwise} \\ h_{l\,i,j}(\mathbf{X}) - y_{l\,i,j} + 1e - 5\sin(\pi t). \end{cases}$$

It follows from Theorem 2.8 that systems (2.4.61) and (2.4.62) achieve GC with respect to the transformation H defined by (2.4.64)–(2.4.66).

Next, select the following initial conditions:

$$(x_{l\,i,j}(0))_{25\times 25} = \mathbf{X}_0(l) + 0.02(rand(25, 25) - 0.5),$$
$$l = 1, 2, 3,$$

$$\boldsymbol{Y}(0) = \boldsymbol{X}(0) + 0.02(rand(25, 25) - 0.5), \tag{2.4.67}$$

where

$$\mathbf{X}_0 = [0.41379 \;\; 0.027763 \;\; -0.012759]^{\mathrm{T}}.$$

The chaotic trajectories of the components $x_{k\,13,12}$ and $y_{k\,13,12}$, $x_{k\,13,13}$ and $y_{k\,13,13}$, as well as $x_{k\,13,14}$ and $y_{k\,13,14}$, of the state variables $\mathbf{X}$ and $\mathbf{Y}$ over the time interval [0, 20], are shown in Figs. 2.4.26–2.4.28, which shows clear chaotic behaviors.

The chaotic orbits of some components $x_{l\,i,j}$, $y_{l\,i,j}$ and of the state variables $\mathbf{X}$ and $\mathbf{Y}$ are shown in Figs. 2.4.29(a)–(f). It can be observed that the dynamic behaviors of the neighboring cells at the lattice: (13, 12), (13, 13), and (13, 14) are quite different.

Figures 2.4.30(a)–(c) show that, although there are initial perturbations (2.4.67), the state variables $\mathbf{X}_{l,13,13}$ and $\mathbf{Y}_{l,13,13}$ achieve GC rapidly.

The three-dimensional views of the evolution of the Lorenz CNN at different times are shown in Fig. 2.4.31, in which chaotic waves can been seen clearly. It can also be observed that the irregular chaotic waves shown by the first three columns in Fig. 2.4.31 have been transformed to wall-shaped chaotic waveforms shown by the last three columns in Fig. 2.4.31.

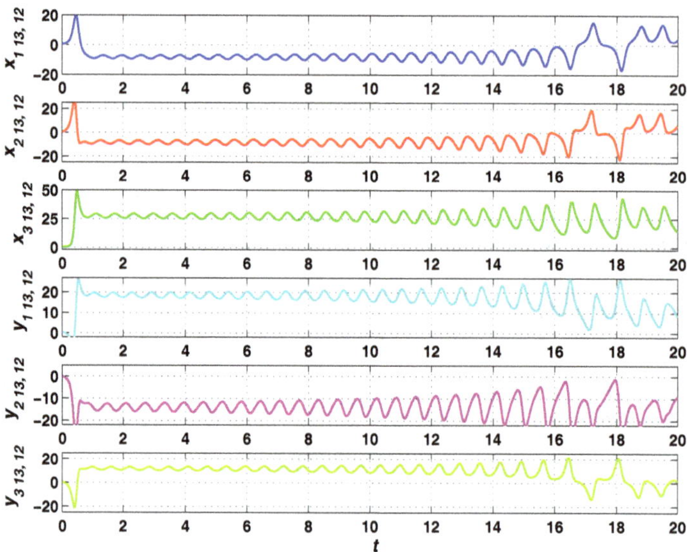

Fig. 2.4.26: Evolution of state variables: $t - x_{1\,13,12}$, $t - x_{2\,13,12}$, $t - x_{3\,13,12}$, $t - y_{1\,13,12}$, $t - y_{2\,13,12}$, and $t - y_{3\,1312}$.

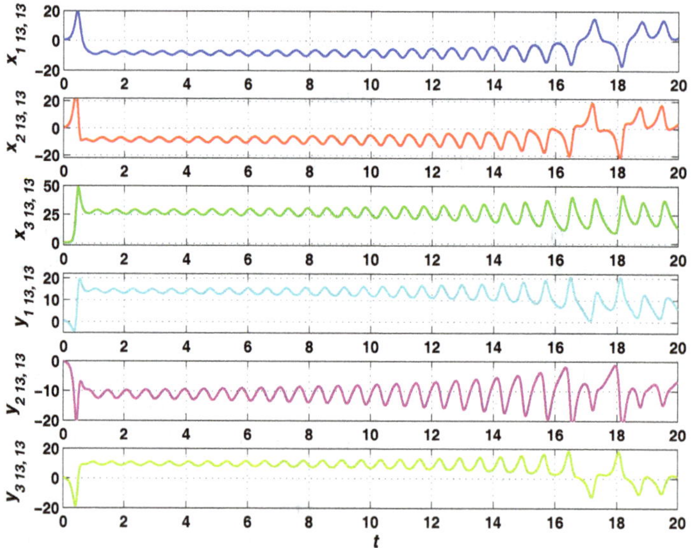

Fig. 2.4.27: Evolution of state variables: $t - x_{1\,13,13}$, $t - x_{2\,13,13}$, $t - x_{3\,13,13}$, $t - y_{1\,13,13}$, $t - y_{2\,13,13}$, and $t - y_{3\,1313}$.

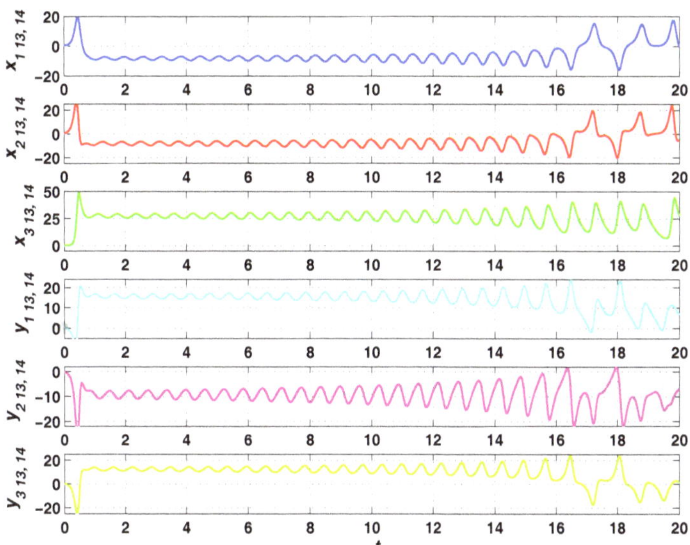

Fig. 2.4.28: Evolution of state variables: $t - x_{1\,13,14}$, $t - x_{2\,13,14}$, $t - x_{3\,13,14}$, $t - y_{1\,13,14}$,$t - y_{2\,13,14}$, and $t - y_{3\,1314}$.

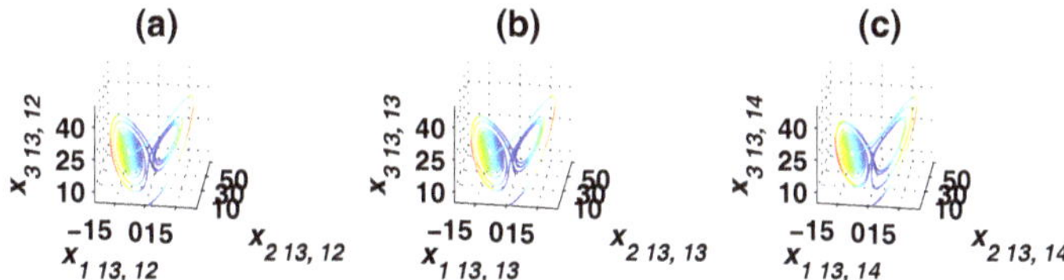

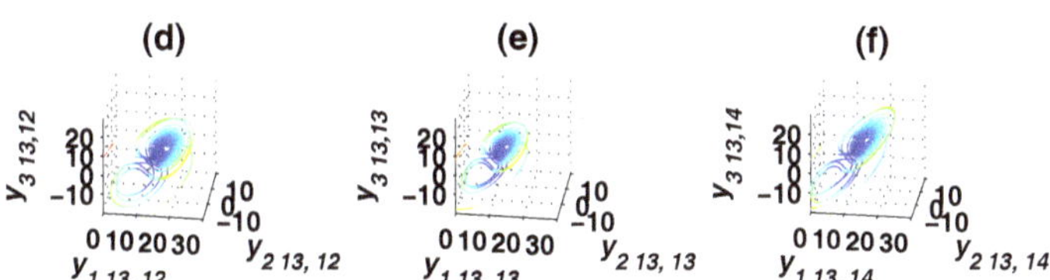

Fig. 2.4.29: Chaotic trajectories of some components of the state variables: (a) $x_{1\,13,12} - x_{2\,13,12} - x_{3\,13,12}$, (b) $x_{1\,13,13} - x_{2\,13,13} - x_{3\,13,13}$, (c) $x_{1\,13,14} - x_{2\,13,14} - x_{3\,13,14}$, (d) $y_{1\,13,12} - y_{2\,13,12} - y_{3\,13,12}$, (e) $y_{1\,13,13} - y_{2\,13,13} - y_{3\,13,13}$, (f) $y_{1\,13,14} - y_{2\,13,14} - y_{3\,13,14}$.

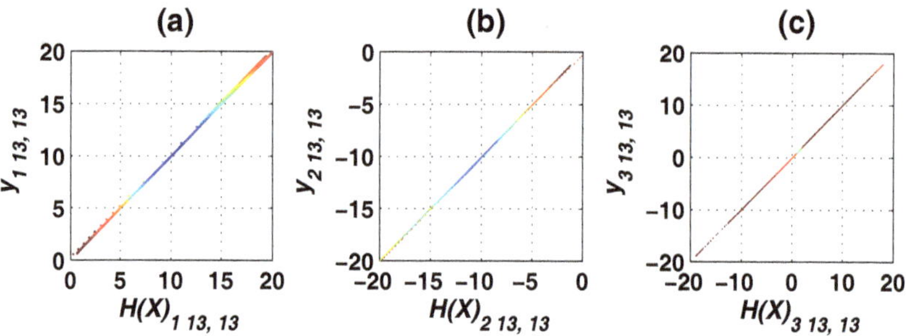

Fig. 2.4.30: (a), (b) and (c) show that the variables $x_{1\,13,13}, x_{2\,13,13}, x_{3\,13,13}$, and $y_{1\,13,13}, y_{2\,13,13}, y_{3\,13,13}$ are in chaos GC with respect to a transformation H defined by (2.4.64)-(2.4.66).

2.4.6 *Application of PGC Theorem to Non-autonomous Bidirectional CCADS*

This subsection shows an application of Theorem 2.10 to non-autonomous bidirectional CCADS.

Based on the hyperchaotic Chen system [Li *et al.* (2011)], a non-autonomous bidirectional Chen CNN with eight state variables in $8 \times 25 \times 25$ dimensions is constructed.

First, construct a continuously differentiable transformation H as follows:

$$H = T \circ \boldsymbol{B} \circ \tilde{H} : \mathbb{R}^{3\times 25\times 25} \times \mathbb{R}^+ \to \mathbb{R}^{3\times 25\times 25} \tag{2.4.68}$$

where

$$\tilde{H} = (\tilde{h}_1, \tilde{h}_2, \tilde{h}_3) : \mathbb{R}^{3\times 21\times 21} \to \mathbb{R}^{3\times 25\times 25}, \tag{2.4.69}$$

$$\boldsymbol{B} : \mathbb{R}^3 \to \mathbb{R}^3, \tag{2.4.70}$$

and

$$T : \mathbb{R}^{3\times 25\times 25} \times \mathbb{R}^+ \to \mathbb{R}^{3\times 25\times 25} \times \mathbb{R}^+ \tag{2.4.71}$$

such that

$$\begin{aligned} \tilde{h}_1((x_{1\,i,j})_{25\times 25}) &= (\alpha^1_{i,j})_{25\times 25}(x_{1\,i,j})_{25\times 25} \\ &= (\tilde{x}_{1\,i,j})_{25\times 25}, \end{aligned} \tag{2.4.72}$$

$$\begin{aligned} \tilde{h}_2((x_{2\,i,j})_{25\times 25}) &= (\alpha^2_{i,j})_{25\times 25}(x_{2\,i,j})_{25\times 25} \\ &= (\tilde{x}_{2\,i,j})_{25\times 25}, \end{aligned} \tag{2.4.73}$$

$$\begin{aligned} \tilde{h}_3((x_{3\,i,j})_{25\times 25}) &= (\alpha^3_{i,j})_{25\times 25}(x_{3\,i,j})_{25\times 25} \\ &= (\tilde{x}_{3\,i,j})_{25\times 25}, \end{aligned} \tag{2.4.74}$$

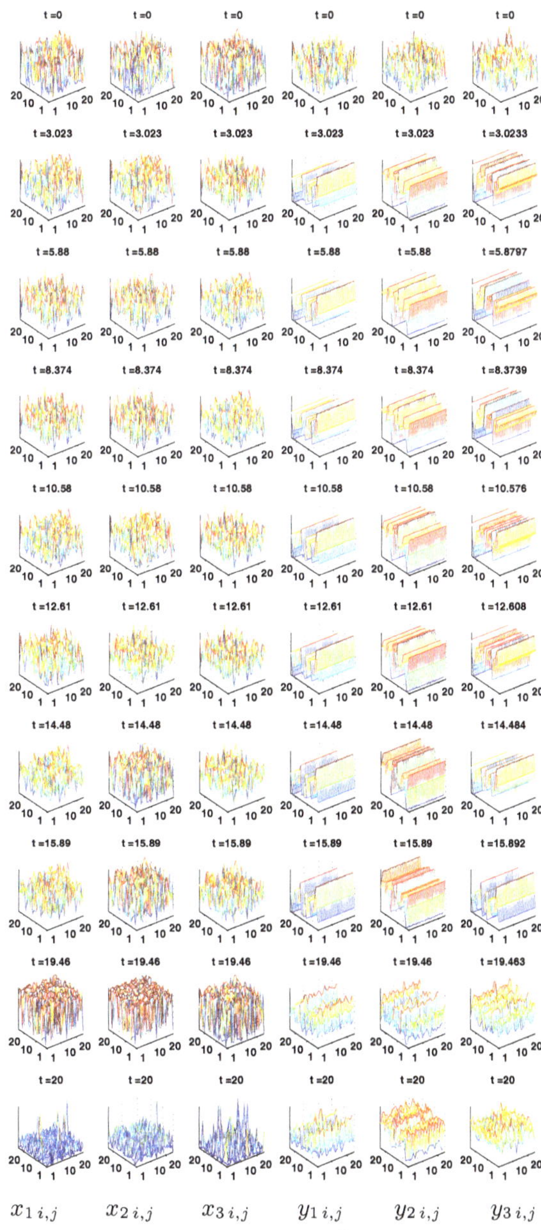

Fig. 2.4.31: The three-dimensional views of the non-autonomous bidirectional Lorenz CNN spiral waves at different time t. The vertical axes represent the state variables $x_{1\,i,j}$, $x_{2\,i,j}$, $x_{3\,i,j}$, $y_{1\,i,j}$, $y_{2\,i,j}$, $y_{3\,i,j}$, while the horizontal axes are the plane coordinates (i, j).

where $\boldsymbol{A}_1 = (\alpha_{i,j}^1)$, $\boldsymbol{A}_2 = (\alpha_{i,j}^2)$, and $\boldsymbol{A}_3 = (\alpha_{i,j}^3)$ are selected to the same as the ones defined by (2.4.37), (2.4.38) and (2.4.39), respectively.

For any triple $(\tilde{x}_{1\,i,j}, \tilde{x}_{2\,i,j}, \tilde{x}_{2\,i,j})$,

$$\begin{aligned}\boldsymbol{B}(\tilde{x}_{1\,i,j}, \tilde{x}_{2\,i,j}, \tilde{x}_{3\,i,j}) &= (\beta_{i,j})_{3\times 3}\tilde{\mathbf{X}} \\ &= \begin{bmatrix} 1\,m & 0 & 0 \\ -1 & -1 & 1 \\ 1 & 1 & 1 \end{bmatrix} \begin{bmatrix} \tilde{x}_{1\,i,j} \\ \tilde{x}_{2\,i,j} \\ \tilde{x}_{3\,i,j} \end{bmatrix}\end{aligned} \tag{2.4.75}$$

and, for any pair $(\boldsymbol{X}, t) \in \mathbb{R}^{3\times 25\times 25} \times \mathbb{R}^+$, one has

$$T(\boldsymbol{X}, t) = \boldsymbol{X} + 0.00005\sin(\pi t). \tag{2.4.76}$$

This transformation $T \circ \boldsymbol{B} \circ \tilde{H}$ is similar to the one shown in Fig. 2.4.1, but it has a term depending on time t.

Next, based on Theorem 2.10, the transformation defined by (2.4.68)-(2.4.76) and the hyperchaotic Chen system [Li *et al.* (2011)] are used to construct a PGC non-autonomous bidirectional CCADS as follows.

The first part of the Chen CNN has the following form:

$$\begin{cases} \dot{x}_{1i,j} = a(x_{2i,j} - x_{1i,j}) + x_{4i,j} \\ \dot{x}_{2i,j} = dx_{1i,j} + cx_{2i,j} - x_{1i,j}x_{3i,j} \\ \dot{x}_{3i,j} = x_{1i,j}x_{2i,j} - bx_{3i,j} \\ \dot{x}_{4i,j} = x_{2i,j}x_{3i,j} + rx_{4i,j} + D[-0.2x_{4i+1,j} \\ \qquad -0.2x_{4i-1,j} - 0.2x_{4i,j+1} - 0.2x_{4i,j-1} + x_{4i,j}], \\ \qquad +0.1sin(\pi t) + D\sin(\pi x_{5i,j}x_{6i,j}x_{7i,j}) \end{cases} \tag{2.4.77}$$

$$i, j = 1, 2, \ldots, 25,$$

where $a = 35; b = 3; c = 12; d = 7; r = 0.5; D = 5 \times 10^{-5}$.

In a compact form, the CCADS (2.4.77) can be written as

$$\dot{\mathbf{X}} = F(\mathbf{X}, \mathbf{Y}, t). \tag{2.4.78}$$

The second part of the CCDAS has the following form:

$$\dot{\mathbf{Y}} = G(\mathbf{Y}, \mathbf{X}, t), \tag{2.4.79}$$

where

$$
\begin{aligned}
g_{l\,i,j}(\mathbf{Y},\mathbf{X},t) &= \sum_{l'=1}^{3}\sum_{i'=1}^{25}\sum_{j'=1}^{25}\frac{\partial h_{l\,i,j}(\mathbf{X}_3,t)}{\partial x_{l'\,i',j'}} \\
&\quad \times f_{l'\,i',j'}(\mathbf{X},\mathbf{Y},t)+\frac{\partial h_{l\,i,j}(\mathbf{X}_3,t)}{\partial t}-q_{l\,i,j}(\mathbf{X},\mathbf{Y},t), \\
&= \sum_{l'=1}^{3}\beta_{l,l'}\sum_{k=1}^{25}\alpha_{i,k}^{l'}f_{l'k,j}(\mathbf{X},Y,t) \\
&\quad +5\times 10^{-5}\pi\cos(\pi t)-q_{l\,i,j}(\mathbf{X},\mathbf{Y},t), \qquad (2.4.80)\\
&\quad l=1,2,3, i,j=1,2,\cdots,25,
\end{aligned}
$$

$$
\begin{aligned}
g_{4\,i,j}(\mathbf{Y},\mathbf{X},t) &= -x_{4\,i,j}, (q_{l\,i,j}(\mathbf{X},\mathbf{Y},t))_{l,25\times 25} \\
&= C((e_{l\,i,j}(\mathbf{X},\mathbf{Y},t))_{l,25\times 25}, \qquad (2.4.81)
\end{aligned}
$$

$$
C=(c_{i,j})_{25\times 25}, \qquad (2.4.82)
$$

$c_{1,1}=0, c_{24,24}=0$, $c_{24,25}=-1$, $c_{25,24}=1$, $c_{25,25}=0$. For other i,j :

$$
c_{i,j}=\begin{cases}-1 & \text{if } i=j,\\ 0 & \text{otherwise}\end{cases} \qquad (2.4.83)
$$

It follows from Theorem 2.10 that systems (2.4.78) and (2.4.79) achieve PGC with respect to the transformation H defined by (2.4.68) – (2.4.76).

Then, select the following initial conditions:

$(x_{l\,i,j}(0))_{25\times 25}=\mathbf{X}_0(l)+0.02(rand(25,25)-0.5), l=1,2,3,4$

where

$$
\begin{aligned}
\mathbf{X}_0 &= [0.46947\ \ 2.3126\ \ 18.285\ \ -34.244]^{\mathrm{T}}, \\
\mathbf{Y}(0) &= \mathbf{X}_0+0.02(rand(25,25)-0.5). \qquad (2.4.84)
\end{aligned}
$$

The chaotic trajectories of the components $x_{k\,13,12}$ and $y_{k\,13,12}$, $x_{k\,13,13}$ and $y_{k\,13,13}$, as well as $x_{k\,13,14}$ and $y_{k\,13,14}$, of the state variables $\mathbf{X}$ and $\mathbf{Y}$ over the time interval [0, 20], are shown in Figs. 2.4.32–2.4.34, which shows clear chaotic behaviors.

Figures 2.4.35(a)–(f) show that, although there are initial perturbations (2.4.84), the state variables $\mathbf{X}_{l,13,13}$ and $\mathbf{Y}_{l,13,13}$ achieve PGC finally. The chaotic orbits of some components $x_{l\,i,j}$, $y_{l\,i,j}$ and of the state variables $\mathbf{X}$ and $\mathbf{Y}$ are shown in Figs. 2.4.36(a)–(f). It can be observed that the dynamic behaviors of the neighboring cells at the lattice: (13, 12), (13, 13), and (13, 14) are quite different.

Three-dimensional views of the evolution of the Chen CNN at different times are shown in Fig. 2.5.37, in which chaotic waves can been seen clearly.

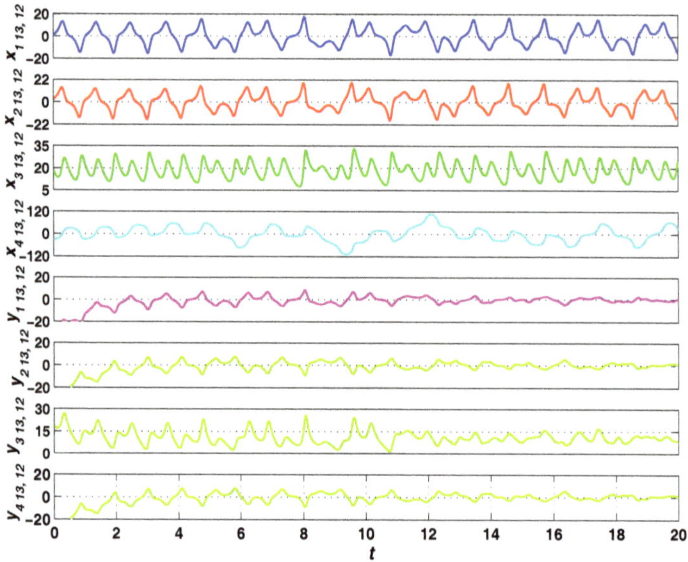

Fig. 2.4.32: Evolution of state variables: $t-x_{1\,13,12}, t-x_{2\,13,12}$, $t-x_{3\,13,12}$, $t-x_{4\,13,12}$, $t-y_{1\,13,12}, t-y_{2\,13,12}$, and $t-y_{3\,1312}$, and $t-y_{4\,13,12}$.

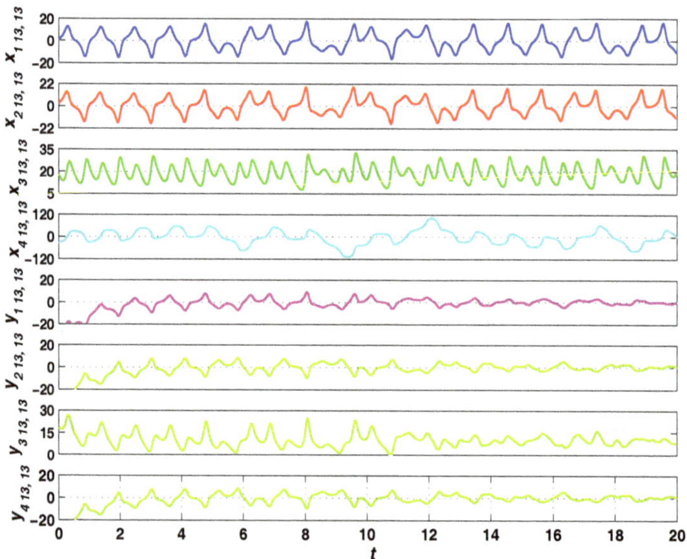

Fig. 2.4.33: Evolution of state variables: $t-x_{1\,13,13}, t-x_{2\,13,13}$, $t-x_{3\,13,13}$, $t-x_{4\,13,13}$, $t-y_{1\,13,13}, t-y_{2\,13,13}$, and $t-y_{3\,1313}$, and $t-y_{4\,13,13}$.

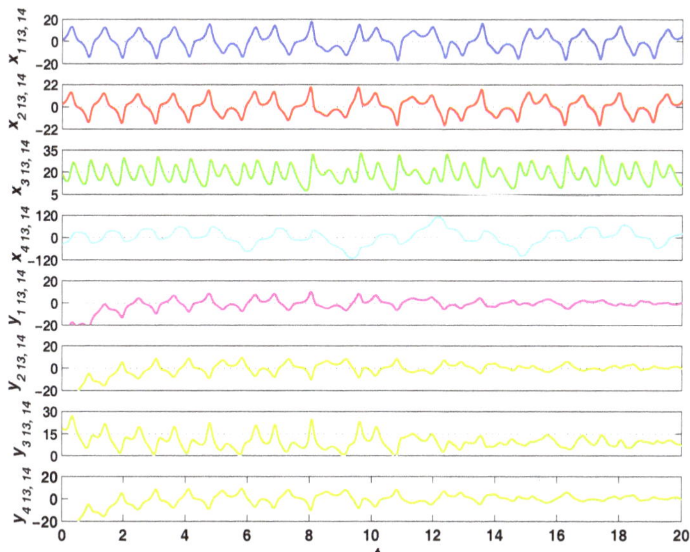

Fig. 2.4.34: Evolution of state variables: $t - x_{1\,13,14}$, $t - x_{2\,13,14}$, $t - x_{3\,13,14}$, $t - y_{1\,13,14}$, $t - y_{2\,13,14}$, and $t - y_{3\,1314}$.

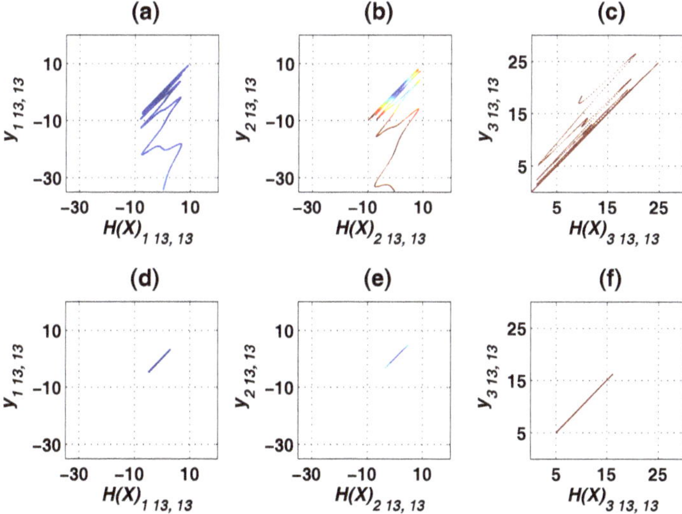

Fig. 2.4.35: (a), (b) and (c) show that the variables $x_{1\,13,13}$, $x_{2\,13,13}$, $x_{3\,13,13}$, and $y_{1\,13,13}$, $y_{2\,13,13}$, $y_{3\,13,13}$ are in chaos GC with respect to a transformation H defined by (2.4.68)–(2.4.76). (d), (e) and (f) show the relationships after $t = 16$.

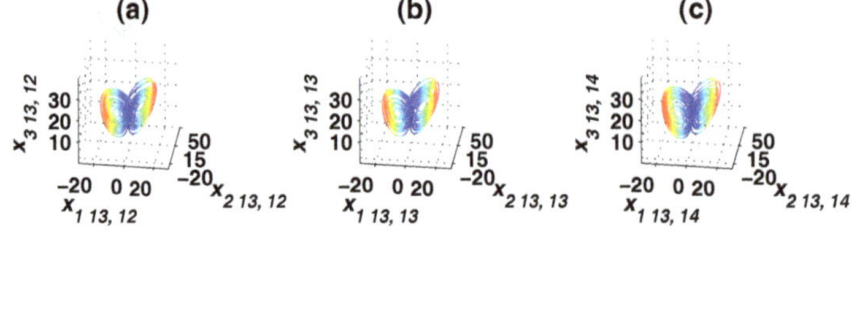

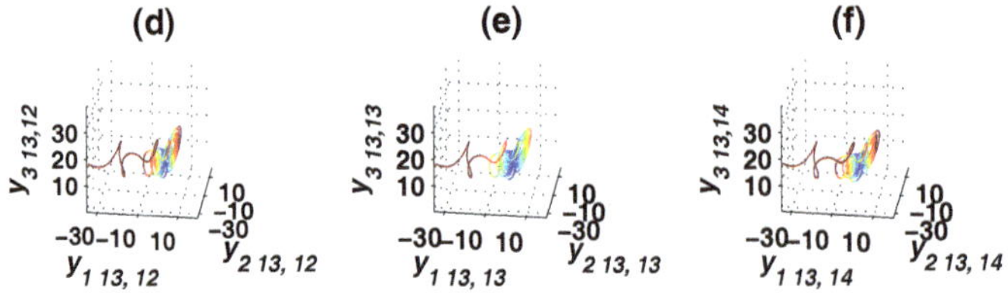

Fig. 2.4.36: Chaotic trajectories of some components of the state variables: (a) $x_{1\,13,12} - x_{2\,13,12} - x_{3\,13,12}$, (b) $x_{1\,13,13} - x_{2\,13,13} - x_{3\,13,13}$, (c) $x_{1\,13,14} - x_{2\,13,14} - x_{3\,13,14}$, (d) $y_{1\,13,12} - y_{2\,13,12} - y_{3\,13,12}$, (e) $y_{1\,13,13} - y_{2\,13,13} - y_{3\,13,13}$, (f) $y_{1\,13,14} - y_{2\,13,14} - y_{3\,13,14}$.

It can also be observed that the irregular chaotic waves shown by the first three columns in Fig. 2.5.37 have been transformed to wall-shaped chaotic waveforms shown by the last three columns in Fig. 2.5.37.

2.5 Conclusions

The main results presented in this part include ten constructive GC theorems, which describe some general forms of two autonomous or non-autonomous CDADS and CCADS to be in GC with respect to a transformation. These theorems have generalized some existing results on GC of vector difference and differential systems [Wang *et al.* (2015a,b); Yang *et al.* (2015); Zhang *et al.* (2015a,b)], and the results on GC of CDADS and CCADS [Min and Zang (2009); Zang *et al.* (2012); Min and Chen (2013)].

The new theorems confirm some equivalent classes of GC systems. The difference between two GC systems in each equivalent class, respectively, is a function that makes the zero solution of the error equation be stable.

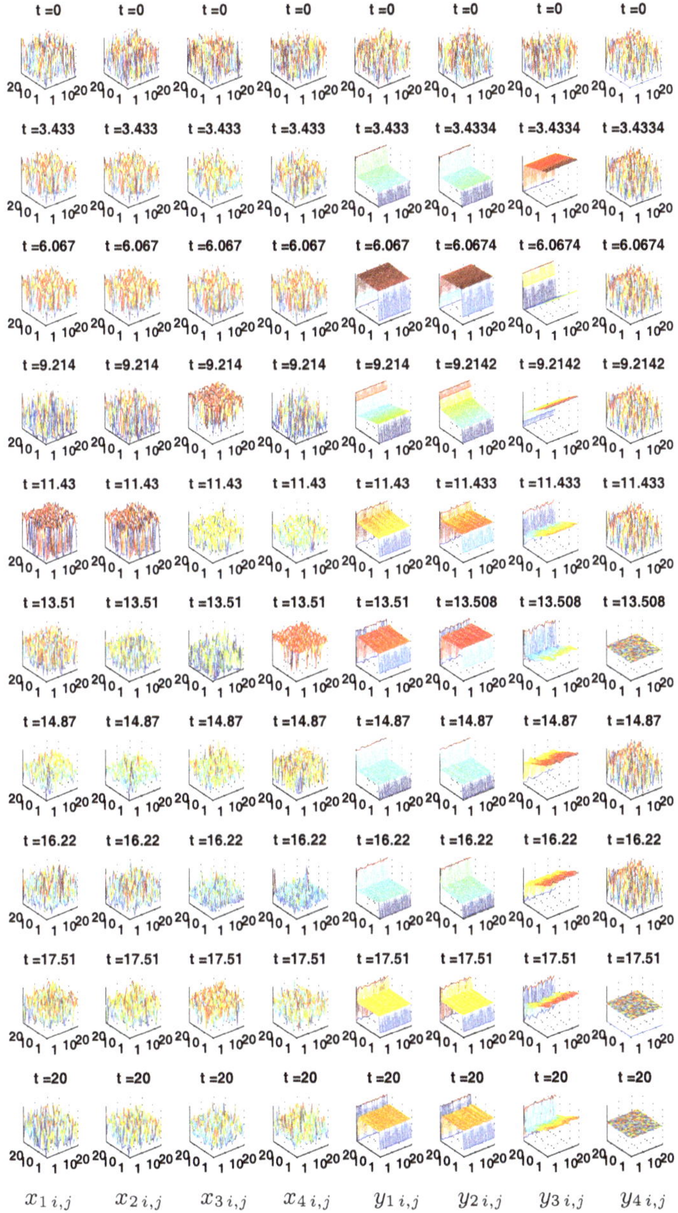

Fig. 2.5.37: The three-dimensional views of the non-autonomous bidirectional Lorenz CNN spiral waves at different time t. The vertical axes represent the state variables $x_{1\,i,j}$, $x_{2\,i,j}$, $x_{3\,i,j}$, $x_{4\,i,j}$, $y_{1\,i,j}$, $y_{2\,i,j}$, $y_{3\,i,j}$, and $y_{4\,i,j}$, while the horizontal axes are the plane coordinates (i, j).

The new approach developed in this part of the book is to use the GC theorems to construct GC CDADS and GC CCADS, rather than to use the Lyapunov exponent techniques as commonly adopted in the literature. Generally speaking, Lyapunov exponents of CDADS and CCADS are very difficult to calculate; therefore, the approach developed in this part of the book should be welcome.

As applications, three GC CDADS and three GC CCADS examples are simulated and analyzed. The simulation results show that the theorems can be easily applied and verified.

Based on the above ten new GC theorems on CDADS and CCADS and the six examples, one can better understand why so many different systems can achieve GC in applications.

Chapter 3

Conclusions

Most phenomena observable from the real world are nonlinear, and the dynamic evolutions of most manmade complex systems are also nonlinear. As a matter of fact, complex dynamic systems are frequently used to model practical problems in diverse disciplines including life science [Fiscus and Fath (2018)], biology science [Bassel (2019)], neuroscience [Barbey (2018)], artificial intelligence [Mabrook *et al.* (2019)], physics [Yang and Perdikaris (2019)], chemistry [Zayed *et al.* (2009)], computer science [Enokido *et al.* (2015)], economics [Leviäkangas and Öörni (2020)], evolutionary computation [Hossain *et al.* (2019)], earth science [Jena *et al.* (2020)], meteorology [Nanda *et al.* (2019)], psychology [Fuster-Parra *et al.* (2015)], sociology [Stadtfeld *et al.* (2020)], and so on.

The cellular neural networks (CNNs), introduced by Chua and Yang [Chua (1997, 1999)], are such typical complex systems. Mathematically, a CNN consists of a spatially discrete collection of continuous nonlinear systems, called cells, which are locally coupled under a coupling law. CNNs have been used for artificial vision, video compression, data fusion, motion and pattern recognition, among others, which are widely observed in both inanimate media and biological media etc.

The studies of complex systems provide new approaches to science that investigates how relationships between parts give rise to the collective behaviors of the whole system and how the system interacts and forms relationships with its environment [Yanee (2002)].

The studies of complex systems have grown dramatically both in volume and in impact. Given the rapid pace of the development, the subject has become more and more diffused. Therefore, there is a need to provide some frameworks for organizing the diversity of mathematical methodologies and approaches. Since most complex systems have some massive interconnected components, network models and basic theories provide important and useful tools for the studies of various complex systems. Within this context, network synchronization is an important inherent subject associated with connected complex systems, which composes the main content of this booklet.

In Chapter 1, synchronization of a coupled discrete array of difference systems (CDADS) and a coupled continuous array of differential systems (CCADS) are studied, which represent most of networked complex systems. In order to describe the internal relationships in CDADS or in CCADS, they are classified into different kinds, respectively; that is, the non-directional arrays of CDADS/CCADS, bidirectional arrays of CDADS/CCADS, non-autonomous arrays of CDADS/CCAS, non-autonomous and bidirectional discrete arrays of CDADS/CCAS. Both partial generalized synchronization of CDADS/CCADS and partial generalized consensus of CDADS/CCADS are investigated in detail.

Since the complexity theory is rooted in chaos theory, the synchronization of two chaotic systems is therefore very important in both theoretical research and practical applications. Furthermore, generalized synchronization (GS), namely the dynamical behaviors of two different complex systems tend to each other with respect to a transformation starting from different initial conditions in a specific domain, is important therefore is studied in detail. In fact, the GS theory can help discover and explain the dynamic relationships between two very different complex systems.

It should be mentioned that, in the past, research on GS theory is not popular due to its methodological difficulties. The question is that, if two systems can achieve GS with respect to a transformation, what kind of representations should the two systems have? In Chapter 1, up to five constructive GS theorems are established for five kinds of CDADS and five kinds of CCADS, respectively. These theorems provide effective frameworks for the ten kinds of GS systems. The theorems also confirm some equivalent classes of GS systems. The differences between two GS systems in each equivalent class is a function that makes the zero solution of the error equation be asymptotically stable. Consequently, it is easy to understand why different systems can generate GS phenomena. The examples given in Section 1.4 help the readers in better understanding the theories and designing transformations in their needs regarding GS of complex systems.

On the other hand, the concept of generalized consensus (GC) of CDADS and CCADS is rooted in both theoretical analysis and practical applications. By nature, this concept is an extension of the system stability and GS. Practically, all computer simulated dynamics of GS system have GC characteristics. For example, if the computed errors for a space matrix $(x_{l\,i,j})_{m\times M\times N}$ are $(\delta_{l\,i,j})_{m\times M\times N}$, then the GS condition given in Definition 1.5, namely,

$$\lim_{k\to+\infty} \|H(\mathbf{X}_m(k)) - \mathbf{Y}(k)\| = 0,$$

should become

$$\lim_{k\to+\infty} \|H(\mathbf{X}_m(k)) - \mathbf{Y}(k)\| = \|(\delta_{l\,i,j})_{m\times M\times N}\|.$$

Indeed, if the computational precision can be small enough, the above formula satisfies the GC definition (2.1).

In Chapter 2, five constructive GC theorems are established, addressing five kinds of CDADS and five kinds of CCADS, respectively. These theorems provide precise representations for the ten kinds of GC systems. Also, these theorems confirm some equivalent classes of GC systems. The differences between two GC systems in each equivalent class is a function that makes the zero solution of the error equation be stable. Consequently, it is easy to understand why different systems can generate various GC phenomena. The examples given in Section 2.4 help the readers in better understanding the theories and designing transformations in their needs regarding GS of complex systems.

In general, a pair of causal events may have some GS or GC relationships. In particular, GC may reveal some relationships for seemingly irrelevant dynamic behaviors generated by causal events. Therefore, it is expected that the theories developed in this booklet can be applied to a wide range of complex systems for in-depth understanding of their dynamic behaviors, overcoming the major difficulties in modeling complex systems, especially in the higher-dimensional settings.

Bibliography

Antal, B., Meusburger, A., and P. Suarsana, E. (2014). *Learning Organizations* (Springer, Learning Organizations).

Barbey, A. K. (2018). Network neuroscience theory of human intelligence, *Trends in Cognitive Sciences* **22**, 1, pp. 8–20.

Bassel, G. W. (2019). Multicellular systems biology: Quantifying cellular patterning and function in plant organs using network science, *Molecular Plant* **12**, 6, pp. 731–742.

Belotti, M., Božić, N., Pujolle, G., and Secci, S. (2019). A vademecum on blockchain technologies: when, which, and how, *IEEE Commun. Surv. Tutor* **21**, 4, pp. 3796–3838.

Brewer, B. B., Carley, K. M., Benham-Hutchins, M., Effken, J. A., and Reminga, J. (2020). Exploring the stability of communication network metrics in a dynamic nursing context, *Social Networks* **61**, 5, pp. 11–19.

Chee, C. Y. and Xu, D. L. (2006). Chaotic encryption using discrete-time synchronous chaos, *Phys. Lett. A* **38**, pp. 284–292.

Chen, E., Min, L. Q., and Chen, G. (2017). Discrete chaotic systems with one-line equilibria and their application to image encryption, *Int. J. Bifurcat. Chaos* **27**, 3, pp. 1750046: 1–17.

Chen, G. and Dong, X. (1998). *From Chaos to Order: Methodologies, Perspectives and Applications (reprinted, 2014)* (World Scientific).

Chen, G. and Uets, T. (1999). Yet another chaotic attractor, *Int. J. Bifurcation and Chaos* **9**, 7, pp. 1465–1466.

Chen, H. K. (2005). Global chaos synchronization of new chaotic systems via nonlinear control, *Chaos Solition Fract.* **23**, 4, pp. 1245–1251.

Chlebus, B. S., Cholvi, V., and Kowalski, D. R. (2020). Universal stability in multi-hop radio networks, *Journal of Computer and System Sciences* **114**, Dec., pp. 48–64.

Chua, L. O. (1997). CNN: A version of complexity, *Int. J. Bifurcation and Chaos* **7**, 10, pp. 2219–2425.

Chua, L. O. (1999). Passivity and complexity, *IEEE Transactions on Circuits and Systems I: Fundamental Theory and Appplications* **46**, 1, pp. 71–82.

Chua, L. O., Wu, C. W., Huang, A. S., and Zhong, G. (1994). A universal circuit for studying and generating chaos –part I and part II, *IEEE Trans. Circ. Syst.-I: Fund Th Appl.* **49**, pp. 732–761.

Cui, G., Zhuang, G., and Lu, J. (2016). Neural-network-based distributed adaptive synchronization for nonlinear multi-agent systems in pure-feedback form, *Neurocomputing* **218**, 12, pp. 234–241.

de Oliveira, M. T., Reis, L. H. A., Medeiros, D. S., Carrano, R. C., Olabarriaga, S. D., and Mattos, D. M. (2020). Blockchain reputation-based consensus: A scalable and resilient mechanism for distributed mistrusting applications, *Computer Networks* **17**, pp. 107367: 1–16.

Enokido, T., Barolli, L., and Takizawa, M. (2015). Journal of computer and system sciences special issue on reliability and optimization for wireless networking and cloud computing, *Journal of Computer and System Sciences* **81**, 8, pp. 1415–1416.

Fiscus, D. A. and Fath, B. D. (2018). *Chapter 6: Life science lessons from ecological networks and systems ecology in Foundations for Sustainability* (Academic Press).

Fuster-Parra, P., García-Mas, A., Ponseti, F. J., and Leo, F. M. (2015). Team performance and collective efficacy in the dynamic psychology of competitive team: A bayesian network analysis, *Human Movement Science* **49**, April, pp. 98–118.

Gámez-Guzmán, L., Cruz-Hernández, C., López-Gutiérrez, R. M., and Garicia-Guerrero, E. E. (2009). Synchronization of Chua's circuits with multi-scroll attractors: Application to communication, *Commun. Nonlinear Sci.* **14**, pp. 2765–2775.

Gao, H. J., Lam, J., and Chen, G. (2006). New criteria for synchronization stability of general complex dynamical networks with coupling delays, *Phys. Lett. A* **360**, 2, pp. 263–273.

Ge, Z. M. and Lin, G. H. (2007). The complete, lag and anticipated synchronization of a bldcm chaotic system, *Chaos Solition Fract.* **34**, 3, pp. 740–764.

Gross, N., Kinzel, W., Kanter, I., Rosenbluh, M., and Khaykovich, L. (2006). Synchronization of mutually versus unidirectionally coupled chaotic semiconductor lasers, *Optics Communications* **267**, 2, pp. 464–468.

Hirche, S. and Hara, S. (2008). Hirche, in *Proceedings of the 17th World Congress The International Federation of Automatic Control* (IFAC, Seoul, Korea), pp. 8780–8784.

Hossain, M. M., Alam, S., and Delahaye, D. (2019). An evolutionary computational framework for capacity-safety trade-off in an air transportation network, *Chinese Journal of Aeronautics* **32**, 4, pp. 999–1010.

Hu, D. and Cao, H. (2016). Stability and synchronization of coupled rulkov ma-based neurons with chemical synapses, *Communications in Nolinear Science and Numerical Simulation* **35**, June, pp. 105–122.

Hunt, B. R., Ott, E., and York, J. A. (1997). Differentiable generalized synchronization of chaos, *Phys. Rev. E* **54**, 4, pp. 4029–4034.

Imai, Y., Murakawa, H., and Imoto, T. (2003). Chaos synchronization characteristics in erbium-doped fiber laser systems, *Optics Commun.* **217**, 1–6, pp. 415–420.

Jena, R., Pradhan, B., Beydoun, G., Nizamuddin, and Affan, M. (2020). Integrated model for earthquake risk assessment using neural network and analytic hierarchy process: Aceh Province, Indonesia , *Geoscience Frontiers* **11**, 2, pp. 613–634.

Ji, Y., Liu, T., and Min, L. (2008). Generalized chaos synchronization theorems for bidirectional differential equation and discrete systems with applications, *Phys. Lett. A* **372**, pp. 3645–3652.

Jia, L., Dai, H., and Meng, H. (2010). A new four-dimensional hyperchaotic Chen system and its generalized synchronization, *Chinese Phys.* **19**, 10, pp. 125–135.

Lago-Fermandez, L. F., Huerta, F., Corbacho, F., and Siguenza, J. A. (2000). Fast response and temporal coherent oscillations in small-world networks, *Phys. Rev. Lett.* **84**, pp. 2758–2761.

Lau, F. C. M. and Tse, C. K. (2003). *Chaos-based Digital Communication Systems* (Springer).

Leviäkangas, P. and Öörni, R. (2020). From business models to value networks and business ecosystems - what does it mean for the economics and governance of the transport system, *Utilities Policy* **64**, June, pp. 101046:1–9.

Li, N., Pan, W., and L. Yan et al. (2014). Enhanced chaos synchronization and communication in cascade-coupled semiconductor ring lasers, *Communications in Nonlinear Science and Numerical Simulation* **19**, 6, pp. 1874–1883.

Li, X. and Chen, G. (2003). Complexity and synchronization of the world trade web, *Physica A* **328**, pp. 287–296.

Li, Y., k. S. Tang, W., and Chen, G. (2011). Generating hyperchaos via state feedback control, *Int. J. Bifurcat. Chaos* **15**, 10, pp. 3367–3375.

Liu, T., Ji, Y., Min, L., Zhao, G., and Qin, X. (2007). Generalized synchronization theorem for non-autonomous differential equation with application in encryption scheme, in *Proc. of the International Conference on Computational Intelligence and Security* (published by the IEEE Computer Society Press, Harbin, China), pp. 972–976.

Liu, T., Min, L., and Cao, L. (2010). Generalized synchronization of non-autonomously discrete time chaotic system, in *Proceeding of 2010 Int. Conf. on Communications, Circuits and Systems*, Vol. II (IEEE Press, Chengdu, China), pp. 786–790.

Liu, Z. and Chen, G. (2003). On a possible mechanism of the brain for responding to dynamical features extracted from input signals, *Chaos Soliton Fract.* **18**, 4, pp. 785–794.

Lü, J. and Chen, G. (2002). A new chaotic attractor couned, *Int. J. Bifurcat. Chaos* **12**, 3, pp. 659–661.

Mabrook, M. M., Khalil, H. A., and Hussein, A. I. (2019). Artificial intelligence based cooperative spectrum sensing algorithm for cognitive radio networks, *Procedia Computer Science* **163**, pp. 19–29.

Min, L. and Chen, G. (2013). Generalized synchronization in an array of nonlinear dynamic systems with applications to chaotic CNN, *Int. J. Bifurcat. Chaos* **23**, 1, pp. 1350016: 1–53.

Min, L. and Chen, G. (2017). Generalized stability in an array of nonlinear dynamic systems with applications to chaotic CNN, *Int. J. Bifurcat. Chaos* **27**, 2, pp. 175002: 1–46.

Min, L., Crounse, K. R., and Chua, L. O. (2000). Analytical criteria for local activity of reaction-diffusion CNN with four state variables and applications to the hodgkin-huxley equation, *Int. J. Bifurcat. Chaos* **10**, pp. 1295–1343.

Min, L. and Zang, H. (2009). Generalized chaos synchronization theorem for array differential equations with application, in *Proc. of 2009 Int. Conf. on Communications, Circuits and Systems*, Vol. I (Chengdu, China), pp. 599–604.

Murali, K. and Laskshmanan, M. (1998). Secure communication using a compound signal from generalized synchronizable systems, *Phys. Lett A* **241**, pp. 303–310.

Na, C., Scheiterer, R. L., Obradovic, D., and Nossek, J. A. (2009). A kalman filter approach to clock synchronization of cascaded network elements, in *Proceedings of the First IFAC Workshop on Estimation and Control of Networked Systems* (Venice, Italy), pp. 24–26.

Nanda, T., Sahoo, B., and Chatterjee, C. (2019). Enhancing real-time streamflow forecasts with wavelet-neural network based error-updating schemes and ecmwf meteorological predictions in variable infiltration capacity model, *Journal of Hydrology* **575**, Aug., pp. 890–910.

Panitz, R. and Glückler, J. (2020). Network stability in organizational flux: The case of in-house management consulting, *Social Networks* **61**, pp. 170–180.

Pecora, L. M. and Carroll, T. L. (1990). Synchronization in chaotic systems, *Phys. Rev. Lett.* **64**, 8, pp. 821–824.

Pei, L., Dai, X., and Li, B. (1997). Chaotic synchronizaon system and electrocardiogram, *Commun. Nonlinear Sci.* **2**, 1, pp. 17–22.

Samli, R., Senan, S., Yucel, E., and Orman, Z. (2019). Stability of air flows in mine ventilation networks, *Neural Networks* **116**, 8, pp. 198–207.

Sausedo-Solorio, J. M. and Pisarchik, A. (2014). Synchronization of map-based neurons with memory and synaptic delay, *Physics Letters A* **378**, 30, pp. 2108–2112.

Semin, M. A. and Levin, L. Y. (2019). Stability of air flows in mine ventilation networks , *Process Safety and Environmental Protection* **12**, 4, pp. 167–171.

Shahverdiev, E. M. (2019). Effect of parameter mismatches on chaos synchronization between josephson junctions coupled in series and driven by a central junction, *Physica C: Superconductivity and its Applications* **561**, 15, pp. 52–57.

Sparrow, C. (1982). *The Lorenz Equations: Bifurcations. Chaos, and Strange Attractors* (Springer-Verlag, NY).

Sprot, J. (2003). *Chaos and Time-Series Analysis* (Oxford University Press, Oxford, UK).

Stadtfeld, C., Takács, K., and Vörös, A. (2020). The emergence and stability of groups in social networks, *Social Networks* **60**, Jan., pp. 129–145.

Subbiah, R., Iyengar, S. S., Radhakrishnan, S., and Kashyap, R. (1993). An optimal distributed algorithm for recognizing mesh-connected networks, *Theoretical Computer Science* **120**, pp. 261–278.

Takaba, K. (2011). Robust synchronization of multiple agents with uncertain dynamics, in *Proceedings of the 18th World Congress The International Federation of Automatic Control* (IFAC, Milano, Italy), pp. 8780–8784.

Wang, X. and Chen, G. (2002a). Synchronization in scale-free dynamical networks: Robustness and fragility, *IEEE Trans. Circ. Syst.-I* **49**, pp. 54–62.

Wang, X. and Chen, G. (2002b). Synchronization in small-world dynamical networks, *Int. J. Bifurcat. Chaos* **12**, pp. 187–192.

Wang, X., Min, L., and Chen, E. (2015a). A generalized stability theorem for discrete chaos systems with application in avalanche image encryption, in *2015 4th International Conference on Information Technology and Management Innovation* (Shenzhen, China), pp. 1021–1034.

Wang, X., Min, L., and Zhang, M. (2015b). A generalized stability theorem for continuous chaos system and design of pesudorandom number generator, in *2015 International Conference on Computational Intelligence and Security* (Shenzhen, China), pp. 375–380.

Wu, C. H. (2008). *Synchronization in Complex Networks of Nonlinear Dynamical Systems* (World Scientific Pub. Co., Singapore).

Wu, C. W. and Chua, L. O. (1993). Transmission of digital signals by chaotic synchronization, *Int. J. Bifurcat. Chaos* **3**, 6, pp. 1619–1627.

Wu, C. W. and Chua, L. O. (1995). Synchronization in an array of linearly coupled dynamical systems, *IEEE Trans. Circ. Syst.-I: Fund Th Appl.* **42**, pp. 430–447.

Wu, L. and Zhu, S. (2003). Multi-channel communication using chaotic synchronization of multi mode lasers, *Phys. Lett. A* **308**, 2–3, pp. 157–161.

Yanee, B.-Y. (2002). General features of complex systems, *in Encyclopedia of Life Support Systems* (EOLSS UNESCO, Oxford, UK).

Yang, T. and Chua, L. O. (1996). Channe-independent chaotic secure communication, *Int. J. Bifurcat. Chaos* **328**, pp. 2653–2660.

Yang, T. and Chua, L. O. (1999). Generalized synchronization of chaos via linear transformations, *Int. J. Bifurcat. Chaos* **9**, 1, pp. 215–219.

Yang, X., Min, L., and Zhang, M. (2015). Generalized stability theorems for bidirectional discrete systems and differential equations with application, *Int. J. of Modeling and Optimization* **5**, 4, pp. 257–267.

Yang, Y. and Perdikaris, P. (2019). Adversarial uncertainty quantification in physics-informed neural networks, *Journal of Computational Physics* **394**, Oct., pp. 136–152.

Zang, H. and Min, L. (2008). An image encryption scheme based on generalized synchronization theorem for discrete array systems, in *Proc. of the 2008 Int. Conf. on Communications, Circuits and Systems*, Vol. II (IEEE Press, Xiamen, China), pp. 948–953.

Zang, H., Min, L., and Zhao, G. (2007). A generalized synchronization theorem for discrete-time chaos system with application in data encryption scheme, in *Proc. of the 2007 Int. Conf. on Communications, Circuits and Systems*, Vol. II(2007) (Kokura, Fukuoka, Japan), pp. 1325–1329.

Zang, H., Min, L., and Zhao, G. (2012). Generalized chaos synchronization theorem for bidirectional array differential and discrete systems with applications, in *the Abstract Book of the 5th Chaotic Modeling and Simulation*

International Conference (The 54th CMSIC, Athens, Greece), pp. 167–168.

Zang, H., Min, L., and Zhao, G. (2013). Generalized chaos synchronization of bidirectional discrete systems, *Chinese Phys. Lett.* **30**, pp. 040502: 1–4.

Zayed, J. M., Nouvel, N., Rauwald, U., and Scherman, O. A. (2009). Chemical complexity: Supramolecular self-assembly of synthetic and biological building blocks in water, *Chemical Society Reviews* **39**, pp. 2806–2816.

Zhang, M., Min, L., and Yang, X. (2015a). Generalized stability theorems for non-autonomous differential equations with application, in *2015 International Conference on Cyber-Enabled Distributed Computing and Knowledge Discovery* (Xian, China), pp. 156–163.

Zhang, M., Wang, D., Min, L., and Wang, X. (2015b). A generalized stability theorem for discrete-time nonautonomous chaos system with applications, *Mathematical Problems in Engineering* **2015**, pp. 121359: 1–12.

Zhang, W. and Zhuang, X. (2019). The stability of chinese stock network and its mechanism, *Physica A: Statistical Mechanics and its Applications* **515**, pp. 748–761.

Zhang, X. and Min, L. (2000). Theory for constructing generalized synchronization and applications, *J. Univ. Sci. Technol. B.* **7**, 3, pp. 225–228.

Index

www.ingramcontent.com/pod-product-compliance
Lightning Source LLC
Chambersburg PA
CBHW050630190326
41458CB00008B/2214

* 9 7 8 9 8 1 1 2 1 4 2 7 1 *